Graphene Composite Supercapacitor Electrodes

David J. Fisher

Published by **Materials Research Forum LLC**
Millersville, PA 17551, USA

Published as part of the book series
Materials Research Foundations
Volume 124 (2022)
ISSN 2471-8890 (Print)
ISSN 2471-8904 (Online)

Print ISBN 978-1-64490-192-2
ePDF ISBN 978-1-64490-193-9

Distributed worldwide by

Materials Research Forum LLC
105 Springdale Lane
Millersville, PA 17551
USA
http://www.mrforum.com

Printed in the United States of America
10 9 8 7 6 5 4 3 2 1

Table of Contents

Batteries are ubiquitous and power everything from the smallest hand-held device to powerful motor vehicles. Ever-increasing numbers of batteries, especially those for vehicles, are based upon lithium. The move away from gasoline-powered transport helps to reduce greenhouse gases but is making lithium yet another strategic material in the same way that rare earths have become. A rival to batteries in certain applications is the capacitor. These provide rapid bursts of back-up energy but generally cannot store as much power as a battery does. So-called supercapacitors, also sometimes known as ultracapacitors and electrical double-layer capacitors, have however increasingly begun to rival batteries in that respect. They also have the advantage of being able to be made very physically flexible.

One fundamental change has been the increasing ability to manipulate the nanoscale structure of carbon-based supercapacitors. This advance obviously offers the secondary advantage of sequestering increasing amounts of carbon from the environment and thus again potentially helping to limit global warming.

A useful means of comparing batteries and supercapacitors is the Ragone plot (figure 1), which maps the relationship between energy and power; essentially a guide therefore to the rate of discharge. The inability of batteries to discharge efficiently at high rates is one of their drawbacks. In the case of lead-acid batteries, the glaring disadvantage is their high density. A second problem is their lifetime. There tends to be a direct relationship between the number of cycles and the depth-of-discharge per cycle. If a lead-acid battery is routinely depleted by 80% rather than 20%, its life may be reduced by as much as 90%.

The fundamental basis of any capacitor is of course that it consists of a pair of electrodes which is surrounded by an electrolyte. When a potential difference exists between the electrodes, oppositely charged ions of the electrolyte are attracted to an electrode and remain there even after the potential difference no longer exists. When the electrodes are later connected via some device, electron flow removes the charge-difference between the electrodes and provides useable energy.

Carbon has long been the favored electrode material because of its good conductivity and its ability to form low-density porous structures. The smaller the pores, the larger the surface area and theoretically the greater the quantity of charge which can be retained. It was originally supposed that, if the pores were too small, ions would be unable to enter them; thus restricting the storage ability. It was later found[1] that this assumption was ill-conceived, and that ions were in fact not impeded from entering even the smallest pores. This discovery has rejuvenated the field of capacitor design, and especially the use of graphene to create supercapacitors.

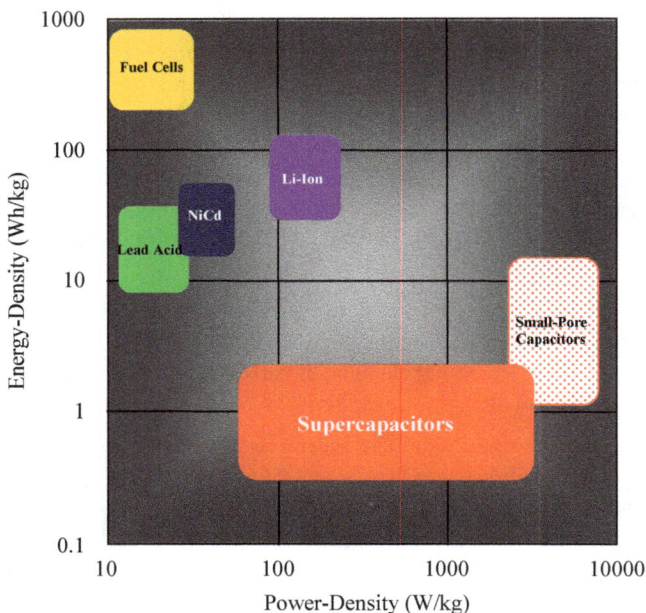

Figure 1. Ragone diagram for supercapacitors and various batteries

Batteries expend the energy stored in chemical bonds while supercapacitors exploit the electrostatic separation between ions and electrodes of high surface area. This leads to capacitances of the order of some tens of Farads per gram. Traditional dielectric capacitors offer capacitances which are only of the order of microfarads. The storage ability of supercapacitors is directly proportional to the capacitance of the electrodes, and this in turn arises from the nm-sized separations between electrolyte ions and carbon and from the high specific surface area of carbon electrodes. The latter is obviously a function of the pore size. Mesoporous carbon, synthesized by using template techniques, can produce controlled pore sizes as small as 2nm; perhaps too small. Carbon nanotubes provide large pores and a high conductivity; increasing the power-density, but decreasing the energy-density. Carbide-derived carbons offer the host carbide lattice as a tailorable template, leading to a high specific capacitance.

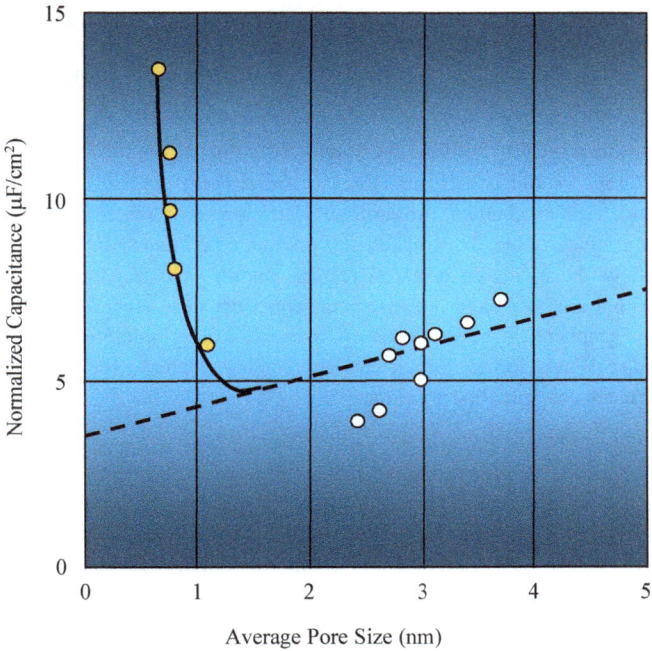

Figure 2. Capacitance of graphene electrode as a function of pore-size. Dotted line: curve-fitting and extrapolation based upon previous results. Solid line: new results.

Although supercapacitors can have ratings of thousands of Farads the energy which is stored is relatively modest, because they are also rated at only about 2.5V per capacitor. The energy-densities may be less than 10Wh/kg. The solution would be to increase that voltage, given that the energy stored in a capacitor depends upon the square of the voltage, but the minute distances between the charge-bearing surfaces mean that the intervening medium could easily break down.

So because supercapacitors store energy by forming a double layer of electrolyte ions on the surface of a conductive electrode they are not limited by the electrochemical charge-transfer kinetics of a battery and can operate at very high charge- and discharge-rates, and can have lifetimes of over a million cycles. On the other hand the amount of energy which is stored in a supercapacitor is an order-of-magnitude smaller than that stored in a battery. The energy-density of a supercapacitor based upon porous activated carbon is of the order of 5Wh/kg, while that of a lead-acid battery can be up to 34Wh/kg. One research aim is therefore to increase the energy-density without impairing the cycle-life or reducing the power-density.

Graphene has a theoretical specific surface area of $2630m^2/g$, thus offering a potentially high energy-storage capacity. It also possesses a very high intrinsic electrical conductivity and high mechanical strength. It is of low cost when produced in large quantities from graphite oxide. This makes it of phenomenal interest and the object in this work is to review the available research results, mainly with regard to the effect of modifying the materials which are used to construct the electrodes. A principal disadvantage of graphene is nevertheless an irreversible agglomeration, due to strong interlayer van der Waals forces, which can appreciably decrease the surface area and result in a poor electrochemical performance.

Environmental Benefits

As alluded to above, the manufacture of supercapacitors can remove carbon from the environment. This can even involve the direct remediation of waste. Because two of the most common materials used in supercapacitors are also the most common ones used in zinc-carbon batteries, recycling of the latter has numerous virtues. Electronic waste is an ever-growing problem and re-use of the carbon and manganese dioxide from dry-cells is possible[2]. The performance of graphene and MnO_2 which had been obtained in this way was tested in 1M H_2SO_4 by using symmetrical (MnO_2|polyvinylalcohol:H_2SO_4|MnO_2) and asymmetrical (graphene|polyvinylalcohol:H_2SO_4|MnO_2) supercapacitors. The asymmetrical device could offer an energy-density as high as 68Wh/kg at a power-density of 8010W/kg, with an operating potential of up to 1.6V and a life of 5000 cycles. This performance was attributed to the large surface area and high conductivity of the graphene and nanoporous MnO_2 which had been recycled from scrap dry cells.

Dry cells are already part of the electronics industry and the associated 'e-waste', but there seems to be no end to the types of scrap which can be turned into the graphene component of supercapacitors. Porous graphene nanosheets can be prepared by microwave-treatment of various forms of biomass. Traditional heating may produce unsuitable amorphous carbon or minute graphite clusters, while microwave treatment of black sesame has produced few-layer porous graphene[3]. As compared with traditional carbonization routes, samples which were prepared by microwave irradiation possessed a larger specific surface area ($2414.5m^2/g$). When used as an electric double-layer capacitor electrode material, the product exhibited a specific capacitance of 369F/g at 1.0A/g, and

320F/g at 10.0A/g. The symmetrical device offered a maximum energy-density of 25.2Wh/kg. A graphene-like-structured nanoporous carbon can be prepared by using jute stick as the carbon precursor[4]. The resultant nanoporous carbon consists of a graphene sheet-like network plus amorphous carbon; with their ratio depending upon the activation temperature. As that temperature is increased, the amorphous carbon is converted into a stable graphene-like network having a specific surface area of $2396m^2/g$, a graphene sheet-like morphology and highly-ordered graphitic sp2 carbon. It exhibits a specific capacitance of 282F/g, with nearly 70% retention at high current rates. Symmetrical test supercapacitors offer an energy-density of 20.6Wh/kg at a power-density of 33600W/kg. Activated-carbon plus graphene composites can be prepared by using cigarette-filter waste as the precursor[5]. Such composites were tested as electrode materials for various electrolytes: ammonium-salt based conventional organic, and two imidazolium-based ionic liquids. This could extend the operating-voltage window up to 3.4V. Electrical double-layer capacitors which were assembled by using the carbon composites could offer specific capacitances of about 160F/g at 0.25A/g; corresponding to energy-densities of some 65Wh/kg at 210W/kg, when ionic liquids were used as the electrolyte. Sulfur-doped porous 3-dimensional graphene materials have similarly been prepared[6] from lignin and polyethersulfone films by using a laser direct writing technique. The film could be transformed into sulfur-doped laser-induced graphene by scribing under ambient conditions. Micro-supercapacitors were directly prepared by using two laser-scribed graphene electrodes and a polyvinylalcohol-H_2SO_4 electrolyte. The devices offered an areal capacitance of $22mF/cm^2$, and an areal energy-density of $1.53mWh/cm^2$ at an areal power-density of $25.4mWh/cm^2$. Graphene-like porous carbon networks possessing 3-dimensional hierarchically ordered so-called ion highways have been obtained[7] by the carbonization and activation of orange-peel waste. They exhibited a specific surface area of $1150m^2/g$.

A carbonized wood graphene oxide and polyvinyl alcohol composite[8] had a specific capacitance of 288F/g, 91% capacitance retention, an energy-density of 36Wh/kg and a power-density of 3600W/kg. Paper-making black-liquor lignin carbon nanosheets, and multilayer graphene derived from Ni^{2+} catalytic graphitization, were used[9] to prepare supercapacitor electrodes. A sample which was sintered at 1000C had a specific capacitance of 163.7F/g at a scanning frequency of 0.2V/s. An asymmetrical supercapacitor offered an energy-density of 26.2Wh/kg at a power-density of 124.6W/kg; with 89.37% capacitance retention after 2000 cycles. Poplar wood pulp was used[10] for the synthesis of a pulp-fiber|reduced-graphene-oxide composite for electrodes. The optimum fiber/graphene ratio was 5:4, and this electrode composition had an areal specific capacitance of $683mF/cm^2$ at $1mA/cm^2$, with 87.5% capacitance retention after 10000

Materials Research Forum LLC

https://doi.org/10.21741/9781644901939

cycles at $5mA/cm^2$. A device which was built by using these electrodes offered a maximum energy-density of $47.71\mu Wh/cm^2$ and a maximum power-density of $1251\mu W/cm^2$. Poplar powder was blended[11] with graphene oxide, and the blended mix was co-doped with B and N. Electrode material for supercapacitors was prepared by 1-step carbonization. The current density reached $0.5A/g$, the specific capacitance exceeded $437F/g$, the specific surface area and the total pore volume reached $1186.1667m^2/g$ and $1.347904cm^3/g$, respectively; with better than 97.8% capacitance retention after 3000 cycles. A resultant symmetrical ultracapacitor offered an energy density of $8.6Wh/kg$ at $498W/kg$ and of $6.9Wh/kg$ at $3096W/kg$.

Flexible graphene supercapacitor electrodes were prepared[12] from biomass-derived carbon dots and graphene composite film and hydrogel, with or without an electrodeposited $NiCo_2S_4$ layer. Dispersed carbon dots with many hetero-atoms inhibited the aggregation of graphene sheets and increased the number of active sites. Composite films with $NiCo_2S_4$ had a specific capacitance of up to $1348F/g$ at a current density of $0.5A/g$. A flexible symmetrical supercapacitor with a 1M H_2SO_4 as an aqueous electrolyte had a specific capacitance of $313F/g$ at $0.5A/g$ and an energy-density of $85.1Wh/kg$ at a power-density of $353W/kg$. When polyvinylalcohol-H_2SO_4 gel was the electrolyte, there was 83.1% capacitance retention after 10000 cycles and 96.8% capacitance retention after 1000 bends.

Tannic acid is a natural polyphenolic biomass which possesses a high electro-activity but negligible electrical conductivity. A plain 3-dimensional porous network, a Cu-modified scale-like microstructure and a Ni-modified flower-like structure were therefore tested[13]. As-prepared plain, Ni-containing, Cu-containing and Fe-containing electrodes had specific capacitances of 373.6, 412.4, 460.4 and $429.4F/g$, respectively, at $1A/g$. Symmetrical supercapacitors which were based upon them offered energy-densities of 14.76, 16.76, 19.13 and $17.6Wh/kg$, respectively, at $300W/kg$.

Three-dimensional graphene structures have been produced by the continuous sequential formation and transformation of glucose-based polymers into foam-like materials[14]. Various activation means then created micro- and meso-pores to form biomass-derived carbon having a Brunauer-Emmett-Teller specific surface area of $3657m^2/g$. The product had a specific capacitance of $175F/g$ in an ionic liquid electrolyte. A supercapacitor which was based upon the material offered a maximum energy-density of $74Wh/kg$ and a maximum power-density of $408kW/kg$.

Reduced graphene oxide was used[15] as a conductive binder for a supercapacitor which was based upon used coffee grounds. A symmetrical 2-electrode system provided high charge storage, with a specific capacitance of $512F/g$ in KOH electrolyte at $0.5A/g$. Other

Materials Research Forum LLC
https://doi.org/10.21741/9781644901939

electrolytes could give a specific capacitance of 440F/g at 0.5A/g. The highest energy-density of 187.3Wh/kg corresponded to power-density of 438W/kg.

A composite was prepared[16] by the simultaneous growth of MnS and reduced graphene oxide on nickel foam, and activated carbon which was based upon walnut shells. A resultant supercapacitor offered an energy-density of 37.9Wh/kg at a power-density of 1500W/kg, and an energy-density of 21.3Wh/kg at a power-density of 3750W/kg.

Graphene nanosheets have been synthesized[17] by carbonizing brown-rice husks, followed by 1-stage KOH-activation. The graphene nanosheets had an ultra-thin crumpled-silk-veil structure with a surface area of about 1225m^2/g and high porosity. A graphene-nanosheet electrode had a specific capacitance of 115F/g at 0.5mA/cm^2 and an energy-density of 36.8Wh/kg at a power-density of 323W/kg; with 88% retention after 2000 cycles.

A 3-dimensional graphene hydrogel, decorated with biomass phytic acid, was prepared[18] by using hydrothermal and freeze-drying methods. The phytic acid molecules were intercalated into the graphene sheets, thus producing materials having a higher specific surface area, lower density and increased compressive strength. When used as an electrode, the material had a specific capacitance of 248.8F/g at 1A/g. A supercapacitor which was based upon such electrodes offered an energy-density of 26.5Wh/kg and a power-density of 5135.1W/kg; with 86.2% retention after 10000 cycles. Graphene sheet-like porous activated carbon was prepared[19] from *Bougainvillea spectabilis*. When used in a supercapacitor, the material had a specific capacitance of 233F/g at a current density of 1.6A/g and an energy-density of 7.2Wh/kg (symmetrical cell).

Nitrogen-doped carbon nanofiber networks with reduced graphene oxide and bacterial-cellulose free-standing paper has been used[20] as a tough and bendable electrode for a supercapacitor. The electrode had an areal capacitance of 2106mF/cm^2 (263F/g) in KOH electrolyte and 2544mF/cm^2 (318F/g) in H$_2$SO$_4$; with essentially 100% retention after 20000 cycles. The tensile strength was 40.7MPa. A resultant symmetrical supercapacitor had an areal capacitance of 810mF/cm^2 in KOH and 920mF/cm^2 in H$_2$SO$_4$; offering an energy-density of 0.11mWh/cm^2 in KOH and 0.29mWh/cm^2 in H$_2$SO$_4$, plus a maximum power-density of 27mW/cm^2 in KOH and 37.5mW/cm^2 in H$_2$SO$_4$.

Nitrogen and sulfur co-doped graphene-like carbon sheets can be made[21] from coir-pith by mechanical activation. The sheets are amorphous, with a defective porous carbon structure, and exhibit a maximum specific capacitance of 247.1F/g at 0.2A/g; with 75.2% retention at 10A/g. The maximum energy-density is 33.6Wh/kg at 0.2A/g, with a maximum power-density of 4220.0W/kg at 10.0A/g. A symmetrical supercapacitor device offered a capacitance of 33.7F/g at 0.2A/g when operated at 1.0V, with 82.0% retention at 10A/g. Few-layer graphene-like nanosheets containing micropores and

Materials Research Forum LLC

https://doi.org/10.21741/9781644901939

mesopores have been produced[22] via the mechanical exfoliation of waste peanut shells. They had a specific surface area of $2070 m^2/g$ and a pore volume of $1.33 cm^3/g$. When incorporated into a supercapacitor, the electrodes exhibited a specific capacity of $186 F/g$ in 1M H_2SO_4 electrolyte. The highest energy-density was 58.125Wh/kg and highest power-density was 37.5W/kg. The working potential increased to 2.5V in an organic electrolyte, leading to an energy-density of 68Wh/kg. As a final example of the exploitation of vegetable matter, cellulose from waste paper has been used[23] to make composites for supercapacitor electrodes. *In situ* polymerization was used to synthesize composites which comprised cellulose, polypyrrole and graphene. The inclusion of cellulose increased the specific capacitance by 318%. In the case of polypyrrole-graphene composites there was a 273% increase in specific capacitance upon adding cellulose. Other bio-waste has been transformed[24] into porous graphene sheets at 900C by using KOH as an activation agent to create porosity and as a catalyst to induce graphitization. The resultant material had a specific surface area of $2308 m^2/g$, a pore volume of $1.3 cm^3/g$, a graphene sheet-like morphology with an interlayer d-spacing of 0.345nm and highly-ordered sp2 carbon. When used as an electrode for a supercapacitor application it led to a specific capacitance of $240 F/g$ at $1A/g$. A typical symmetrical supercapacitor offered 87% capacitance retention at 50A/g, and 93% retention after 25kcycles. The energy-density was 21.37Wh/kg at a power-density of 13420W/kg.

Typical urban waste can also be turned into supercapacitor materials. Scrap plastics can be turned into graphene nanosheets by using bentonite to degrade the plastic in a 2-step pyrolysis process at 450C and 945C under nitrogen[25]. The use of the sheets as supercapacitor electrodes led to a specific capacitance of $398 F/g$, with a scan-rate of 0.005V/s. The supercapacitor exhibited an energy-density and power-density of 38Wh/kg and 1009.74W/kg, respectively. Graphene|mesoporous carbon electrode materials have been made[26] from waste polyethylene plastic by low-temperature carbonization at 700C. The material has a surface area of $1175 m^2/g$ and $2.30 cm^3/g$ of mesopores, with a good electrochemical performance in symmetrical and hybrid supercapacitors with wide voltage windows. A hybrid supercapacitor which had the above material as the anode and $LiMn_2O_4$ as the cathode was operated at 2.0V with an 0.5M Li_2SO_4 electrolyte. The hybrid supercapacitor offered an energy-density of 47.8Wh/kg at a power-density of 250W/kg, with 83.8% retention after 5000 cycles. An energy-density of 63.3Wh/kg was offered by a 4.0V symmetrical supercapacitors when using 1-ethyl-3-methylimidazolium tetrafluoroborate as the electrolyte; with 89.3% retention after 5000 cycles. Spongy nitrogen-doped graphene can be prepared[27] from waste polyethylene-terephthalate bottles by mixing with urea at various temperatures. Nitrogen-fixation affected the structure and morphology of the material, improving charge-propagation and ion diffusion. When used as a supercapacitor electrode the specific capacitance was up to 405F/g at 1A/g, with an energy-density of 68.1Wh/kg and a maximum power-density of 558.5W/kg in 6M KOH electrolyte; with 87.7% capacitance retention after 5000 cycles at 4A/g.

The very most environmentally-sensitive pollutants can also be turned to good use. Nitrogen-enriched graphene-like carbon nanosheet can be produced from bio-oil by using graphitic carbon nitride as a self-sacrificing template and nitrogen source[28]. The resultant material has a nitrogen content of 8.09at%, a thickness of about 2.8nm and an electrical conductivity of 85S/m. When used in a symmetrical supercapacitor, the energy-density is 10.6Wh/kg at a power-density of 131.1W/kg in 1M H_2SO_4. Argon-plasma enhanced chemical vapor deposition can be used[29] to synthesize vertically-oriented graphene from waste oil. This yields large-area material (12cm x 3.5cm) with a highly-oriented structure. When used in a supercapacitor, the electrode exhibits a 4 times higher capacitance than that of conventionally sourced material. Composites with MnO_2 lead to a maximum energy-density of some 33.2Wh/kg at 1.0kW/kg and a power-density of 10.2kW/kg at 22.9Wh/kg. High-yield quantized nitrogen-doped graphene nanodiscs have been prepared[30] from waste tyres under high pressure and temperature. Upon increasing the temperature to between 600 and 1100C, the carbon atoms rearranged so as to build a mixed graphene structure of nanodiscs and quantum dots. When used as an electrode in a supercapacitor, the specific capacitance was 161.24F/g, with a power-density of 733.3W/kg and an energy-density of 27.1Wh/kg. In a composite of asphalt it was found[31] that the introduction of 1% of graphene oxide could increase the conductivity to about 400% and the surface area to some 114%. Composite materials without added conducting material exhibited a high capacitance and cycling stability. Their rate capability was maintained from 0.5 to 100A/g, with about 88% capacitance retention. The energy-density was 22.0Wh/kg at a power-density of 55.4kW/kg.

Choice of Electrodes

Much of the work on supercapacitors revolves around the materials which are used for the electrodes. Even when attention is specifically restricted to graphene supercapacitors, there remains a wide choice with regard to the selection of the other electrode in an asymmetrical device and with regard to modification of the graphene itself. Graphene electrodes are typically prepared from graphene oxide sheets and subjected to hydrazine-gas reduction so as to restore the conductive carbon network. In a typical case[32] a maximum specific capacitance of 205F/g, with a power-density of 10kW/kg at an energy-density of 28.5Wh/kg was found for an aqueous electrolyte solution. Related supercapacitor devices exhibited a long cycle-life along, with some 90% of the specific capacitance being retained after 1200 cycles. Another supercapacitor with graphene-based electrodes exhibited[33] a specific energy-density of 85.6Wh/kg at room temperature and 136Wh/kg at 80C for a current density of 1A/g. Such parameters are comparable to those of Ni metal hydride batteries. An interesting feature was that the intrinsic surface capacitance and specific surface area of single-layer graphene could be fully exploited by using curved graphene sheets that would not re-stack; a major problem with graphene. The curved morphology permitted the formation of mesopores which could be wetted by ionic liquids that could operate at above 4V. A further design[34] involved using chemically reduced graphene oxide as the electrode material and an ionic liquid, with added acetonitrile, as the electrolyte. The specific capacitance, energy-density and specific power-density were 132F/g, 143.7Wh/kg and 2.8kW/kg, respectively. An electrochemical method was used[35] to reduce graphene oxide under constant potential, and a supercapacitor which based upon the films offered a specific capacitance of 128F/g and an energy-density of 17.8Wh/kg; operating within a potential window of 1.0V in a 1M $NaNO_3$ electrolyte. Some 86% of the specific capacitance was retained after 3500 charge-discharge cycles. A wide range of materials can be combined with graphene in order to produce composite electrodes, always with the main object being to maximize the energy-density of supercapacitors.

OXIDES

AgVO₃

Graphene|$AgVO_3$ nanocomposites were synthesized[36] by using a 1-step chemical-bath method which anchored the nanoparticles to the graphene surface. The material had an energy-density of 10Wh/kg at a power-density of 25W/kg, and a power-density of 2045W/kg at an energy-density of 6Wh/kg.

Al₂O₃

A reduced-graphene-oxide|polyindole|γ-Al_2O_3 ternary nanocomposite electrode was used[37] to construct an asymmetrical supercapacitor with a $HClO_4$ solution electrolyte. The composite had a specific capacitance of 308F/g. When combined with reduced graphene oxide as the negative electrode, the asymmetrical device could be cycled at 0 to 1.4V and offered an energy-density of 10Wh/kg at a 3880W/kg power-density, with a specific capacitance of 38.46F/g and 83% capacitance retention after 5000 cycles.

BiFeO₃

A composite of $BiFeO_3$|graphene was coated onto a stainless-steel substrate by drop-casting[38]. A supercapacitor was constructed from the composite and aqueous 1M Na_2SO_4

electrolyte. The device offered a power of 18.75kW/kg and an energy-density of 1.9Wh/kg.

Bi_2O_3

A nano-welding technique, based upon the microwave-heating of graphene, was used[39] to make a carbon-nanotube-supported Bi_2O_3 electrode having a graphene|carbon-dot encapsulated structure. The as-prepared composite electrode had a capacitance of $1.90mAh/cm^2$ at $1mA/cm^2$ and a rate-performance of $1.57mAh/cm^2$ at $100mA/cm^2$. A resultant device which had the composite as an anode, and carbon nanotube loaded nickel cobalt as a cathode, offered a maximum energy-density of $589.3\mu Wh/cm^2$ (98.2Wh/kg) at a power-density of $0.8mW/cm^2$, and $288.3\mu Wh/cm^2$ (48.1Wh/kg) at $57.1mW/cm^2$; with 80.1% capacitance retention after 8000 cycles.

$BiVO_4$

A graphene|$BiVO_4$ free-standing monolith composite was prepared[40] hydrothermally in which flexible graphene sheets acted as a skeleton. When used as a binder-free electrode in a 3-electrode system, the composite electrode has a specific capacitance of 479F/g at a current-density of 5A/g. A symmetrical supercapacitor which was based upon this composite offered an energy-density of 45.69Wh/kg at a power-density of 800W/kg, and rapid delivery of 10.75Wh/kg with a power-density of 40kW/kg.

$CeMo_2O_8$

Nanoparticles of $CeMo_2O_8$ were anchored to the surface of N,P-doped reduced graphene oxide by using a sonochemical approach[41]. The resultant electrode had a specific capacitance of 638F/g at 2mV/s. A symmetrical supercapacitor which was based upon the composite electrodes offered an energy-density of 29.7Wh/kg at 500W/kg and a power-density of 16kW/kg at 14.3Wh/kg; with complete capacitance retention after 4000 cycles at 100mV/s.

CeO_2

Ternary graphene|cerium-oxide|porous-polyaniline nanocomposites were synthesized[42] by polymerization and electrochemical reduction. The electrochemical capacitance of the nanocomposite could attain 454.8F/g at 1.0A/g when the mass-ratio of the oxide and the porous polyaniline was 1:4, with 70-23% capacitance retention after 10000 cycles at a current density of 5.0A/g in 1M H_2SO_4 electrolyte. A graphene-aerogel|cerium-oxide nanoparticle composite, prepared hydrothermally[43], had a specific capacitance of 365F/g and 376.2F/g at 5mV/s and 0.5A/g, respectively, with 91.4% capacitance retention after 4000 cycles.

$CoFe_2O_4$

A composite of $CoFe_2O_4$|reduced-graphene-oxide|polyaniline was used[44] as a negative electrode, while rod-shaped β-$Co(OH)_2$ was used as the positive electrode. The supercapacitor had a linear capacitance of 1.41F/m and an energy-density of 2.70 x 10^{-6}Wh/cm. A ternary electrode of hybrid $CoFe_2O_4$|graphene|polyaniline[45] had a maximum specific capacitance of 1123F/g, an energy-density of 240Wh/kg at 1A/g, a power-density of 2680W/kg at 1A/g and 98.2% capacitance retention after 2000 cycles. The polyaniline increased electron transport, while the ferrite nanoparticles prevented re-stacking of the carbon sheets. An asymmetrical supercapacitor was based[46] upon a core-shell nanocomposite, $CoFe_2O_4$|methylcellulose-carbohydrate-polymer, as the positive electrode and a phenylenediamine-graphene aerogel as the negative electrode in aqueous KOH electrolyte. The specific capacitance was 433.3F/g at a current density of 1A/g. The device offered an energy-density and power-density of about 73Wh/kg and 1056W/kg, respectively, with 89% capacitance retention after 2000 cycles.

$CoMn_2O_4$

An asymmetrical supercapacitor was constructed[47] from a $CoMn_2O_4$-modified graphene nanoribbon electrode and a plain graphene nanoribbon electrode, in Na_2SO_4 electrolyte. When cycled reversibly at 0 to 1.9V, the device offered an energy-density 84.69Wh/kg at a power-density of 22kW/kg; with 96% capacitance retention.

$CoMoO_4$

A hydrothermal method was used[48] to synthesize $CoMoO_4$|graphene composite in which Co^{2+} ions were adsorbed on the graphene oxide by electrostatic interaction and then treated with $(NH_4)_6Mo_7O_{24}$. The composite had a specific capacitance of about 394.5F/g and offered an energy-density of about 54.8Wh/kg at 1mV/s.

Asymmetrical supercapacitors were based[49] upon a reduced graphene oxide negative electrode and a $CoMoO_4$ positive electrode. The graphene nanosheets and $CoMoO_4$ nanostructures had a specific capacitance of about 168.8 and 98.34F/g, respectively. The device had a maximum specific capacitance of 26.16F/g at a current density of $0.5mA/cm^2$ and an energy-density of 8.17Wh/kg, with 85% capacitance retention after 4000 cycles at a current density of $1.0mA/cm^2$.

Pristine $Mn_{1/3}Ni_{1/3}Co_{1/3}MoO_4$ and $Mn_{1/3}Ni_{1/3}Co_{1/3}MoO_4$ dispersed in various percentages of reduced graphene oxide were compared[50]. The $Mn_{1/3}Ni_{1/3}Co_{1/3}MoO_4$|graphene composite had a specific capacitance of 1750F/g at 1A/g in 3M KOH. The $Mn_{1/3}Ni_{1/3}Co_{1/3}MoO_4$|graphene exhibited 85.5% capacitance retention after 5000 cycles at

10A/g. An energy-density of 38.8Wh/kg at a constant power-density of 200W/kg was possible.

Three-dimensional free-standing hierarchical $CoMoO_4|CoS$ core-shell heterostructures on reduced-graphene-oxide|Ni-foam were prepared[51]. The core-shell $CoMoO_4|CoS$ composite had a specific capacitance of 3380.3F/g at 1A/g, with 81.1% capacitance retention after 6000 cycles. An asymmetrical supercapacitor was constructed by using $CoMoO_4|CoS$ and activated-carbon as the positive and negative electrodes, respectively. This device offered an energy-density of 59.2Wh/kg at a power-density of 799.8W/kg; with 91.5% capacitance retention after 6000 cycles.

$CoNiO_2$

A composite of pore-enriched $CoNiO_2$ and reduced graphene oxide hollow fibers was constructed[52] by using wet-spinning and hydrothermal methods. Due to the *in situ* nucleation and growth of porous ultra-thin $CoNiO_2$ nanosheets, and the hollow structure of the reduced graphene oxide fibers, the resultant composite had a specific capacitance of 645.8 and 460.0C/g at 2 and 50A/g, respectively. After cycling for more than 50000 times, there was a 42.9% capacitance increase. A supercapacitor with composite and reduced graphene oxide fiber electrodes offered an energy-density of 43.99Wh/kg at a power-density of 1.70kW/kg, and an energy-density as high as 38.41Wh/kg at a power-density of 21.61kW/kg.

Co_3O_4

An asymmetrical supercapacitor with a mass loading of $10mg/cm^2$ on each planar electrode was constructed[53] by using a graphene|cobaltite nanocomposite as the positive electrode and activated-carbon as the negative electrode. The positive electrode had a capacitance of 618F/g, while those of graphene|Co_3O_4 and graphene|NiO were 340F/g and 375F/g, respectively, under the same conditions. An asymmetrical device with 10mg composite and 30mg activated-carbon offered an energy-density of 19.5Wh/kg, with an operational voltage of 1.4V. The device could offer an energy-density of 7.6Wh/kg at a power-density of about 5.6kW/kg, at a high sweep-rate. The capacitance was retained after up to 10000 cycles.

Nanoparticles (20nm) of Co_3O_4 were grown[54] *in situ* grown onto chemically-reduced graphene oxide sheets for use as a pseudocapacitor electrode in a 2M KOH aqueous electrolyte. The reduced-graphene-oxide|Co_3O_4 electrode had a specific capacitance of 472F/g at a scan-rate of 2mV/s, in a 2-electrode cell, with 82.6% capacitance retention when the scan-rate was increased to 100mV/s. It offered an energy-density of 39.0Wh/kg at a power-density of 8.3kW/kg. A supercapacitor electrode[55], with Co_3O_4 nanoparticles on vertically-aligned graphene nanosheets which were supported by carbon fabric, had a specific capacitance of 3480F/g. A very flexible symmetrical supercapacitor was constructed by using the above material for both electrodes. The device had a capacitance of 580F/g, with 86.2% capacitance retention after 20000 cycles. It offered an energy-density of 80Wh/kg and a power-density of 20kW/kg at 27Wh/kg.

An aqueous-electrolyte asymmetrical supercapacitor was based[56] upon Co_3O_4|reduced-graphene-oxide nanoribbon positive electrodes and graphene-film negative electrodes. Immobilized Co_3O_4 nanoparticles on the nanoribbon combined a high electronic conductivity plus short ion-diffusion paths in a mesoporous structure. The device operated at 1.6V and offered an energy-density of 64.2Wh/kg and a power-density of 8.3kW/kg, some 94% capacitance retention after 2000 cycles. Three-dimensional graphene aerogel with embedded oxide nanoparticles was prepared[57] by using solvothermal methods, plus freeze-drying and heat-treatment. The resultant composite had a specific capacitance of 660F/g at 0.5A/g, with 65.1% retention at 50A/g in a 3-electrode system. An asymmetrical supercapacitor was constructed by using the present material as the cathode and porous carbon as the anode in 6M KOH aqueous electrolyte. Over a voltage range of 0.0 to 1.5V, the device offered an energy-density of 40.65Wh/kg and a power-density of 340W/kg, with 92.92% capacitance retention after 2000 cycles. A 1-step hydrothermal method was used to prepare[58] a 3-dimensional porous oxide|graphene aerogel which had a specific surface area of $127m^2/g$ and a pore-size range of 1 to 110nm. The porous oxide increased the specific capacitance of the composite aerogel to 1512.7F/g at 1A/g. The combination of two aerogel electrodes with LiOH-polyvinylalcohol gel electrolyte created an asymmetrical supercapacitor with an operating voltage of 0 to 1.6V, and 81.5% capacitance retention after 5000 cycles. It offered an energy-density of 68.1Wh/kg at a power-density of 982.9W/kg, and a maximum energy-density of 58.7Wh/kg at a power-density of 1209.4W/kg.

Reduced graphene oxide provides[59] an homogeneous structure and improves electron-transport, while a binary oxide has more active sites than those of nickel oxide or cobalt oxide alone. The specific capacitance of a graphene|mixed-oxide composite was about 1750F/g at 1A/g, with 79% capacitance retention after 10000 cycles at 4A/g. An asymmetrical supercapacitor having this material as the positive electrode, and activated-carbon as the negative electrode, offered an energy-density of 37.7Wh/kg at a power-density of 800W/kg.

A composite of nitrogen-doped reduced graphene oxide and nano-sized Co_3O_4 was prepared[60] by using a sonochemical method. When tested using a 3M KCl aqueous electrolyte, a symmetrical device having the composite as both electrodes, exhibited a specific capacitance of 763F/g at a scan-rate of 2mV/s and an energy-density of 138Wh/kg, with 98.9% retention after 4000 cycles. Honeycomb-like porous Co_3O_4, grown[61] onto a 3-dimensional graphene network on nickel foam, was prepared by using a solution-growth method and annealing. A Co_3O_4|graphene|Ni-foam electrode had a specific capacitance of 321F/g at 1A/g, with 88% capacitance retention after 2000 cycles. In a 2-electrode system, the composite offered a maximum energy-density of 7.5Wh/kg at a power-density of 794W/kg; remaining at 4.1Wh/kg with a power-density of 15kW/kg. A flower-like 3-dimensional nanostructure of Co_3O_4|MnO_2 on nitrogen-doped graphene oxide was produced[62] by thermal reduction at 650C in the presence of ammonia and urea. The composite electrode had a specific capacitance of up to 347F/g at 0.5A/g, and a corresponding energy-density of 34.83Wh/kg.

Graphitized porous fullerene carbon soot has been used as a matrix for the preparation[63] of activated-carbon. Graphene|Co_3O_4 composite with cross-linked porous Co_3O_4 nanofibers on graphene sheet was hydrothermally deposited onto a nickel-foam substrate. The graphene|oxide electrode had a capacitance of 1935F/g at 5A/g, with 83% capacitance retention after 2000 cycles. An asymmetrical supercapacitor which incorporated activated-carbon and graphene|oxide electrodes offered an energy-density of 50.3 to 20.9Wh/kg and a power-density of 786 to 12128W/kg; with 77% capacitance retention after 5000 cycles. A dual-template nanocasting method was used[64] to produce ultra-fine (less than 10nm) Co_3O_4 grains, dispersed on reduced graphene oxide nanosheets. Pores of about 4nm in the Co_3O_4|reduced-graphene-oxide provided more diffusion channels for rapid ion/electron transport. The Co_3O_4|reduced-graphene-oxide electrode had a specific capacitance of 709.1F/g at 1A/g; with 91.2% capacitance retention after 6000 cycles. An asymmetrical device with Co_3O_4|reduced-graphene-oxide and reduced-graphene-oxide electrodes offered an energy-density of 48.2Wh/kg at 750.5W/kg.

A graphene|Co_3O_4|polypyrrole ternary nanocomposite was synthesized[65] by using a 2-step technique in which the surface of a graphene|Co_3O_4 binary nanocomposite was covered with a thin film of polypyrrole. The nanocomposites were cast onto Cu|Cu(OH)$_2$ substrates. The ternary nanocomposite, in a 3-electrode system, had a specific capacitance of 422F/g at a scan-rate of 10mV/s, and 385F/g at a current density of 1A/g in 6M KOH. An energy-density of 13.4Wh/kg was found at a power-density of 250W/kg. A resultant symmetrical supercapacitor cell with an organic electrolyte offered an energy-density of 33.06F/g and a power-density of 18.36Wh/kg at 10mV/s.

Reduced-graphene-oxide on nickel foam[66] with Co_3O_4 nanoflakes had a specific capacitance of 1328C/g at a 2A/g current density, and better stability when compared with that of Co_3O_4 nanoflakes deposited onto bare nickel foam. A resultant asymmetrical supercapacitor had a specific capacitance of 80F/g at a current density of 0.1A/g, with 94.5% capacitance retention after 10000 cycles. It offered an energy-density of 20Wh/kg at a power-density of 1200W/kg. Porous Co_3O_4|reduced-graphene-oxide composites were prepared[67] on nickel foam by electrophoretic deposition, thermal reduction and hydrothermal methods. The form of the Co_3O_4 changed, with increasing molar ratio, from dispersed flower-like clusters of fine nanosheets, to coarse flower-like clusters, to a honeycomb-like structure of fine nanosheets and finally to a honeycomb-like structure covered with a coarse flower-like structure. The specific capacitance of a Co_3O_4|reduced-graphene-oxide deposit with a molar ratio of 2:1 attained 1138.11F/g at a current density of 1A/g, and 800.71F/g at a scanning-rate of 5mV/s; with 80.67% capacitance retention after 5000 cycles. A resultant asymmetrical supercapacitor had a specific capacitance of 108.87F/g at a current density of 1A/g, with a stable operational voltage of 1.45V and an energy-density of 647.05Wh/kg.

Three-dimensional porous-carbon|Co_3O_4 composites were prepared[68] via the pyrolysis of a precursor. They had a specific capacitance of 423F/g at 1A/g, with 85.7% capacitance retention at 10A/g, and 83% capacitance retention after 2000 cycles. A resultant asymmetrical supercapacitor, with composite and activated-carbon electrodes, had a potential window of 1.7V and a maximum energy-density of 21.1Wh/kg, with a power-density of 790W/kg. A 3-dimensional heterostructure of Co_3O_4 and Co_3S_4 on nickel foam, covered in reduced graphene oxide, was prepared[69]. In the Co_3O_4|Co_3S_4 nanocomposite, a nanostructure of Co_3S_4 was constructed from Co_3O_4 nanowalls on reduced-graphene-oxide|Ni-foam. The material had a capacitance of 13.34F/cm^2 (5651.24F/g) at a current density of 6mA/cm^2, as compared with that of Co_3O_4|reduced-graphene-oxide|Ni-foam, with its capacitance of 3.06F/cm^2 (1230.77F/g) at the same current density. The oxide|sulfide electrode offered an energy-density of 85.68Wh/kg, a specific power-density of 6048.03W/kg and 86% capacitance retention after 1000 cycles at a scan-rate of 5mV/s. By using Co_3O_4|Co_3S_4|reduced-graphene-oxide|Ni-foam and activated-carbon as positive and negative electrodes, respectively, an asymmetrical supercapacitor was constructed which offered a specific capacitance of 79.15mF/cm^2 at an applied current density of 1mA/cm^2, and an energy-density of 0.143Wh/kg at a power-density of 5.42W/kg.

Electrodes which had been made[70] from hydrothermally synthesized boron-incorporated reduced graphene oxide, Co_3O_4 and Co_3O_4|B-reduced-graphene-oxide nanocomposites were tested in 2M KOH and NaOH electrolytes. The specific capacitance was always higher in KOH than in NaOH. A Co_3O_4|B-reduced-graphene-oxide electrode had a specific capacitance of 600F/g (270C/g) at 0.1A/g and 454F/g (204C/g) at 10A/g in KOH. A Co_3O_4|B-reduced-graphene-oxide in KOH system exhibited 87.8% capacitance retention after 2000 cycles. A Co_3O_4|B-reduced-graphene-oxide in KOH system had a maximum power-density of 2250W/kg, with an energy-density of 12.77Wh/kg at 10A/g.

Nanocomposites of the form, graphene|NiO and graphene|Co_3O_4, were synthesized[71] by using a co-precipitation method. The as-synthesized materials were drop-cast onto as-grown $Cu(OH)_2$ nanowire arrays on copper substrates. Three-electrode measurements in 6M KOH electrolyte showed that graphene|Co_3O_4 and graphene|NiO had capacitances of 342.6 and 652F/g at a scan-rate of 5mV/s, and 278.5 and 667.58F/g at a current-density of 1A/g, respectively. A power-density of 250W/kg led to energy-densities of 23.17 and 9.7Wh/kg for graphene|NiO and graphene|Co_3O_4, respectively. Microstructured globe artichokes of reduced-graphene-oxide|$Ni_{0.3}Co_{2.7}O_4$ composite on nickel foam were prepared[72] in which the artichoke flower-like morphology was made up of hundreds of self-assembled micropetals, interconnected at the base, so as to form microspheres. The morphology of the globe artichokes increased the stability of composite electrodes. The interconnected structure of the binder-free reduced-graphene-oxide|$Ni_{0.3}Co_{2.7}O_4$ electrode enabled better charge transport, leading to a specific capacitance and areal capacitance of 1624F/g and 2.37F/cm², respectively, at a current density of 2A/g. The specific capacitance increased from 1088F/g to 1728F/g after 7000 cycles. When the power-density was increased from 0.5 to 8kW/kg, the energy-density fell from 56.39Wh/kg to 40Wh/kg.

Self-assembled $Mn_{0.5}Co_{2.5}O_4$ nanofibers were sandwiched between graphene sheets by using a hydrothermal treatment[73]. The composite exhibited 93.8% capacitance retention after 10000 cycles at a current density of 10A/g, and has a specific capacitance of 617 and 434F/g at current densities of 30 and 40A/g, respectively. When an asymmetrical supercapacitor was constructed using composite and activated-carbon electrodes, the device offered an energy-density of 36.8 and 13.6Wh/kg, respectively, at a power-density of 147.3 and 3755.8W/kg.

Self-assembled $Mn_{0.5}Co_{2.5}O_4$ nanofibers were intercalated into graphene frameworks having a mesoporous structure by hydrothermal treatment and annealing[74]. The structure prevented volume changes in the oxide, and impeded graphene-layer re-stacking. The composite had a specific capacity of 950mAh/g after 100 cycles at a current density of 0.2A/g. In a 3-electrode system, it had a specific capacitance of 1926F/g at a current density of 10A/g, and capacitances of 1575 and 1211F/g at current-densities of 30 and 40A/g, respectively. When incorporated into an asymmetrical supercapacitor together with activated-carbon, the device offered energy densities of 64.9 and 14.6Wh/kg at power densities of 75.2 and 3754.3W/kg, respectively.

$CuCo_2O_4$

An electrode was created[75] by anchoring $CuCo_2S_4$ nanoparticles to a graphene aerogel using solvothermal methods. The 21nm sulfide particles shortened ion-diffusion paths, and the graphene aerogel provided a conducting skeleton which prevented sulfide agglomeration. The electrode had a specific capacitance of 668F/g at a current density of 1A/g; with 72% retention when the current was increased to 20A/g. A supercapacitor was constructed by using $CuCo_2S_4$|graphene-aerogel composite as a free-standing electrode. This offered an energy-density of 22Wh/kg and a power-density of 1080W/kg. Polyethylene oxide was combined with 5wt% KOH to create polymer cement electrolytes[76]. Electrolytes which contained 2wt% of polyethylene oxide possessed the optimum balance of compressive strength and ionic conductivity. Supercapacitor devices were constructed from the polymer cement electrolyte and reduced-graphene-oxide|$CuCo_2O_4$ nanowires. One device offered an energy-density of 0.2mWh/cm^2 and an areal capacitance of 439.35mF/cm^2. The other device offered an energy-density of 0.74mWh/cm^2 and an areal capacitance of 407.07mF/cm^2.

$CuMn_2O_4$

A graphene|$CuMn_2O_4$ composite[77] was such that, when the mass ratio was 1:1, the as-prepared material exhibited optimum properties. The capacitance attained 342F/g at a current density of 1.0A/g.

CuO

Reduced-graphene-oxide|CuO|polyaniline nanocomposites were synthesized[78] which had various weight-ratios of polyaniline and fixed weight-ratios of oxide and reduced graphene oxide. In 2M Na_2SO_4, composite with a weight of 300mg of polyaniline, had a specific capacitance of 213.20F/g, an energy-density of 18.95Wh/kg and a power-density of 545.79W/kg; with 97.6% capacitance retention after 5000 cycles. A combination of 1-dimensional oxide nanowires and 2-dimensional graphene sheets was used[79] to create a highly porous electrically conductive 3-dimensional composite. The porous nanostructure, when used as a supercapacitor electrode provided a 1.6 times greater capacitance than that of a graphene-only electrode, with 91.2% retention after 5000 cycles. The electrode had a specific energy-density of 50.6Wh/kg at a power-density of 200W/kg. Nanorods of CuO were grown vertically onto graphene nanosheets by electrostatic attraction[80]. The composite had a specific capacity of 2.51C/cm^2 at 2mA/cm^2. A symmetrical supercapacitor which was constructed from the electrodes offered a maximum energy-density of 38.35Wh/kg at a power-density of 187.5W/kg.

Cu_2O

A graphene|Cu_2O nanocomposite film was fabricated[81] on copper foil by electrodeposition and evaluated in 6M KOH. The electrode had a specific capacitance of 161.31F/g at a scan-rate of 10mV/s, and 124.70F/g at a current-density of 1mA/mg. An asymmetrical supercapacitor was constructed by using the nanocomposite|Cu-foil as the positive electrode and graphene|Cu-foil foil as the negative electrode had a maximum specific capacitance of 11.94F/g and offered an energy-density of 6.63Wh/kg at 10mV/s.

Fe_2O_3

Well-crystallized oxide with a particle-size of 35 to 40nm was combined[82] with graphene. An optimum composite with 49.7wt%Fe_2O_3 had a BET surface area of 215.3m^2/g. When used as a supercapacitor electrode, the capacitance was 2310F/g at 5mV/s and 615F/g at a current density of 100A/g. Monocrystalline Fe_2O_3 nanoparticles were grown[83] directly onto graphene hydrogel as an anode material for supercapacitors. During formation of the composite, flexible graphene sheets, decorated with Fe_2O_3 particles, self-assembled to form interconnected porous microstructures with a high specific surface area. This greatly aided charge and ion transport in the electrode. The graphene|Fe_2O_3 electrode had a specific capacitance of 908F/g at 2A/g, for a potential range of -1.05 to -0.3V; with 69% capacitance retention at 50A/g. Electrodes of Fe_2O_3|graphene-oxide paper were developed[84] for high-density supercapacitors. The specific areal capacitance of the electrode was about 3.08F/cm^2 at a current-density of 5mA/cm^2, with 95% capacitance retention after long-term cycling. An Fe_2O_3 nanowire network, wrapped in graphene, was used[85] as a transparent electrode. The specific areal capacitance was 3.3mF/cm^2 at a scan-rate of 10mV/s. and the power-density was 191.3W/cm^3.

Chemical synthesis under hydrothermal condition resulted[86] in the formation of crystalline α-Fe_2O_3 and poorly crystallized or amorphous Fe_3O_4. The particles had a size of 30 to 50nm. Chemical coupling of N-reduced-graphene-oxide with pseudocapacitive Fe_2O_3-Fe_3O_4 increased the performance of the composite. Asymmetrical supercapacitors used the material as positive or negative electrodes. An asymmetrical device which was based upon using the composite material as the positive electrode and activated carbon as the negative electrode had a specific capacitance of 111.95F/g at 0.8A/g, with an energy-density of 44.93Wh/kg and better than 92% capacitance retention after 10000 cycles. An asymmetrical supercapacitor was based[87] upon an α-Fe_2O_3|reduced-graphene-oxide nanocomposite as the negative electrode and an α-MnS|reduced-graphene-oxide nanocomposite as the positive electrode. The supercapacitor had a potential window of 1.6V when 3M KOH was used as the electrolyte. The maximum specific capacitance of the device was 161.7F/g at a current density of 1A/g, with a maximum energy-density of 57.5Wh/kg at a power-density of 800W/kg.

*Table 1. Summary of Ragone-plot data for supercapacitors
with oxide composite electrodes*

Oxide	Energy-Density (Wh/kg)	Power-Density (W/kg)
Al_2O_3	10	3880
Co_3O_4	7.6	5600
Co_3O_4	39.0	8300
Co_3O_4	80	20000
Co_3O_4	64.2	8300
Co_3O_4	40.65	340
Co_3O_4	68.1	982.9
Co_3O_4	58.7	1209.4
Co_3O_4	37.7	800
Co_3O_4	7.5	794
Co_3O_4	4.1	15000
Co_3O_4	50.3	12128
Co_3O_4	20.9	786
Co_3O_4	48.2	750.5
Co_3O_4	20	1200
Co_3O_4	21.1	790
Co_3O_4	0.143	5.42
Co_3O_4	12.77	2250
$Ni_{0.3}Co_{2.7}O_4$	56.39	580
$Ni_{0.3}Co_{2.7}O_4$	40	8000
$Mn_{0.5}Co_{2.5}O_4$	36.8	147.3
$Mn_{0.5}Co_{2.5}O_4$	13.6	3755.8
$Mn_{0.5}Co_{2.5}O_4$	64.9	75.2
$Mn_{0.5}Co_{2.5}O_4$	14.6	3754.3
CuO	38.35	187.5
Fe_2O_3	57.5	800

Fe_2O_3	43.2	200
Fe_2O_3	16.38	1784
Fe_2O_3	87.05	500
Fe_3O_4	147	150
Fe_3O_4	86	2587
Fe_3O_4	10.4	250
Fe_3O_4	29	5200
Fe_3O_4	47.7	550
Fe_3O_4	120.68	3910
Fe_3O_4	60.8	45200
Fe_3O_4	11.4	25800
Fe_3O_4	44	25000
Fe_3O_4	30.4	2600
Fe_3O_4	14.2	50500
Fe_3O_4	25.2	100
Fe_3O_4	2.08	94000
Fe_3O_4	8.34	47000
Fe_3O_4	7400	13
Fe_3O_4	16	300
Fe_3O_4	49.1	4830
Fe_3O_4	31.5	100
Fe_3O_4	26.1	3981
Fe_3O_4	23.68	7270
Fe_3O_4	26.7	32700
Fe_3O_4	24	400
Fe_3O_4	19	10000
Fe_3O_4	45.4	67800
Fe_3O_4	25.7	229200

Fe_3O_4	23.0	450.8
Fe_3O_4	78.9	284.1
Fe_3O_4	10.7	500
Fe_3O_4	5.67	5110
Fe_3O_4	35.22	200
Fe_3O_4	22.2	101
Fe_3O_4	34.5	12400
Fe_3O_4	18.2	400
Fe_3O_4	88	23200
Fe_3O_4	41.8	200
Fe_3O_4	28.2	899.36
Fe_3O_4	33.9	319.3
Fe_3O_4	7.9	11804
Fe_3O_4	59.2	1480
Fe_3O_4	28.8	57600
Fe_3O_4	23.3	2001
Fe_3O_4	24.2	242
Fe_3O_4	105.3	308.1
Fe_3O_4	42.77	30800
Fe_3O_4	2353	196
Fe_3O_4	71.74	850
Fe_3O_4	21.23	1500
Fe_3O_4	64.3	19800
Fe_3O_4	19.6	351
Fe_3O_4	6235	7990
Fe_3O_4	106.7	1200
Fe_3O_4	259	1008
Fe_3O_4	23.2	119.9

Fe_3O_4	110	550
Mn_3O_4	32	833
Mn_3O_4	47.8	1000
Mn_3O_4	23	600
Mn_3O_4	42.2	500.0
Mn_3O_4	18.5	280
Mn_3O_4	27.92	277.78
MoO_3	75.27	816.67
MoO_4	58.6	801.0
Mo_3O_{10}	72.6	217.7
Mo_3O_{10}	13.3	3993.8
Nb_2O_5	89	125
Nb_2O_5	20	3500
Nb_2O_5	72	479
NiO	27.3	1562.6
NiO	32.5	375
NiO	19.78	7500
NiO	55.6	1598.7
NiO	35.9	749.1
NiO	109.8	800
NiO	5.4	430
RuO_2	6.8	49800
RuO_2	19.7	6800
RuO_2	39.28	128010
RuO_2	16.2	9885
RuO_2	101	2500
RuO_2	13	21000
RuO_2	61.2	183.8

RuO_2	43.8	750
RuO_2	39.1	37500
SnO_2	19.4	9973.26
SnO_2	22	238.3
SnO_2	17.1	5803.3
SnO_2	23.4	253
SnO_2	10.2	3684
SnO_2	29.6	5310.26
TiO_2	9.08	598
TiO_2	14.1	8500
VO_2	126.7	10000
V_2O_5	38.8	455
V_2O_5	96	800
V_2O_5	37.2	3743
V_2O_5	26.22	425
V_2O_5	27.6	3600
V_2O_5	13680	22.8
V_2O_5	39	947
V_2O_5	116	1520
V_2O_5	185.86	37.20.
V_2O_5	54.62	1636.5
V_2O_5	26.7	6000
WO_3	25	6000
WO_3	27.2	752
WO_3	93	500
ZnO	18.14	10000
ZnO	35.32	139.72

An Fe_2O_3|graphene nanocomposite was prepared[88] by using a 1-step hydrothermal method. The composite had a maximum specific area of $91m^2/g$ and a smaller pore size of 10nm as compared with those of pristine Fe_2O_3; with a surface area of $76m^2/g$ and a pore size of 17nm. The composite could have a specific capacitance as high as 315F/g at a discharge current-density of 2A/g; at a current density of 10A/g, the specific capacitance could be 185F/g; with 98% capacitance retention after 2000 cycles. Nanoparticles of Fe_2O_3 were decorated onto graphene and carbon nanotube networks[89]. This electrode material had a specific capacitance of 675.7F/g at 1A/g in 6M KOH aqueous electrolyte. An asymmetrical supercapacitor which had Fe_2O_3|graphene-nanotubes|carbon-nanotube composite as the cathode and sulfurized CoAl layered double hydroxide as the anode, offered an energy-density of 60.3Wh/kg.

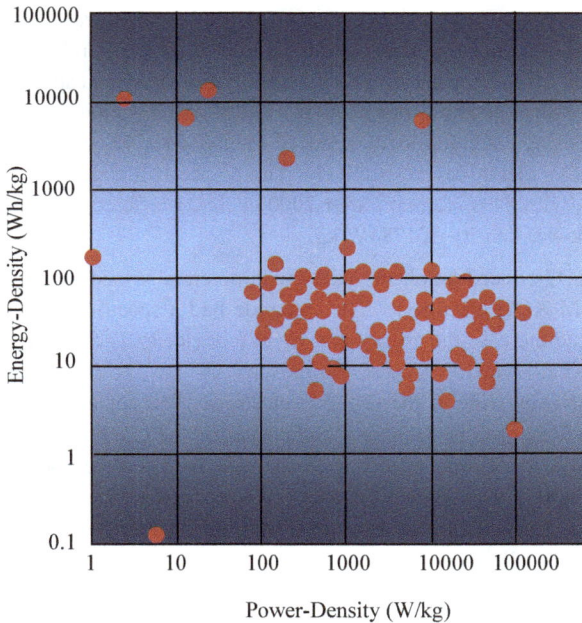

Figure 3. Ragone plot for oxide composite electrodes

Ternary nanostructured polyaniline|Fe_2O_3-decorated-graphene composite hydrogel, coated on carbon cloth, has been studied[90] as an electrode material for flexible supercapacitors. The material had a specific capacitance of 1124F/g at a current density of 0.25A/g in 1M H_2SO_4. When nanoparticles of Fe_2O_3 were grown[91] on graphene|carbon-nanotubes the resultant composites had a specific capacitance of 258F/g at 1A/g. Manganese dioxide was also grown onto the surface of graphene|carbon-nanotubes to form composites. An asymmetrical supercapacitor was constructed by using graphene|carbon-nanotube|Fe_2O_3 composite as the negative electrode and graphene|carbon-nanotube|MnO_2 composite as the positive electrode. The voltage ranged from 0 to 2.0V in 1M Li_2SO_4 electrolyte. The device offered an energy-density of 43.2Wh/kg at 200W/kg.

A 3-dimensional ternary composite of iron oxide embedded polypyrrole|reduced-graphene-oxide matrix was evaluated[92] as an electrode material. The material had a specific capacitance of 442F/g in 1M KCl electrolyte at a current density of 1A/g; with 88% capacitance retention after 8000 cycles. A 2-step method was used[93] to prepare ternary 3-dimensional reduced-graphene-oxide|$Ni_{0.5}Zn_{0.5}Fe_2O_4$|polyindole nanocomposite as an electrode for a supercapacitor. The material had a specific capacitance of about 320F/g at 0.3A/g over a potential range of -0.1 to 1V. An asymmetrical supercapacitor was constructed using 3-dimensional reduced-graphene-oxide and graphene as the negative and positive electrodes, respectively. The device had a specific capacitance of 48.9F/g at 0.5A/g, with 84% retention after 2000 cycles. It offered an energy-density of 16.38Wh/kg at a power-density of 1784W/kg.

Composites of Fe_2O_3|reduced-graphene-oxide, with potential windows of -0.15 to -1.2V, were tested[94] in 6M KOH electrolyte. The electrode had a specific capacity of 413C/g. These features could increase the energy-density of a device to 77.5Wh/kg by combining a modified $La_{0.85}Sr_{0.15}MnO_3$|$NiCo_2O_4$ composite, as the positive electrode, with Fe_2O_3|reduced-graphene-oxide as the negative electrode. The device exhibited a power-density of 54000W/kg. A ternary nanocomposite of Fe^{II} and Fe^{III} oxide|reduced-graphene-oxide|polypyrrole was used[95] as electrodes in a supercapacitor. A specific capacitance of 626.8F/g was found for the Fe_2O_3|reduced-graphene-oxide|polypyrrole nanocomposite at a constant current density of 1A/g in 1M H_2SO^4 electrolyte. The high specific capacitance for Fe_3O_4|reduced-graphene-oxide|polypyrrole) was attributed to the presence of electrical double-layer capacitance and pseudocapacitance mechanisms. The Fe_2O_3 |reduced-graphene-oxide |polypyrrole nanocomposite offered the highest energy-density of 87.05Wh/kg at a power-density of 500W/kg.

Fe_3O_4

Two-dimensional sandwich-like sheets of iron oxide were created[96] on graphene via the direct growth of FeOOH nanorods on the graphene surface and their subsequent electrochemical transformation into Fe_3O_4. The Fe_3O_4|reduced-graphene-oxide nanocomposites had a capacitance of 326F/g and an energy-density of 85Wh/kg in 1M LiOH solution. An Fe_3O_4|graphene nanocomposite was used[97] as the negative electrode, and a graphene-based 3-dimensional porous carbon with a surface area of $3355m^2$/g, was used as the positive electrode in a supercapacitor. The Fe_3O_4|graphene nanocomposite had a reversible specific capacity of more than 1000mAh/g at 90mA/g and 704mAh/g at 2700mA/g. The supercapacitor offered an energy-density of 147Wh/kg at a power-density of 150W/kg, and of 86Wh/kg at a power-density of 2587W/kg. This energy-density was comparable to that of lithium-ion batteries, and the power-density was that of a symmetrical supercapacitor.

Nitrogen-doped graphene quantum dots were deposited[98] onto Fe_3O_4-halloysite nanotubes and used as anode materials for a supercapacitor. The Fe_3O_4 nanoparticles, prepared by co-precipitation, were deposited onto the nanotube surfaces and coated with (3-aminopropyl)-triexthoxysilane in order to anchor the 4 to 10nm quantum-dots by forming amide linkages. The resultant composite had a specific capacitance of 418F/g in neutral electrolyte solutions, with an energy-density of 10.4 to 29Wh/kg and a power-density of 0.25 to 5.2kW/kg. A ternary reduced-graphene-oxide|Fe_3O_4|polyaniline composite was synthesized[99] in the form of 3-dimensional Fe_3O_4-decorated reduced graphene oxide doped polyaniline nanorods. The specific capacitance of the nanocomposite was about 283.4F/g at 1.0A/g, and offered a maximum energy-density of 47.7Wh/kg at a power-density of 550W/kg; with 78% capacitance retention after 5000 cycles.

Hexagonal magnetite nanoplatelets were synthesized[100] hydrothermally on holey graphene nanoribbons and tested in a polyvinylalcohol-H_2SO_4 gel electrolyte. The optimum loading of the nanoribbons was a maximum of 30wt% of magnetite nanoplatelets. At above 50wt% magnetite loading, the structural identity of the nanoribbon was impaired, with increased network resistivity depletion. The mass loading of the magnetite was inversely related to ion diffusion and electronic conduction. Balanced ionic and electronic conduction in the 30wt% magnetite nanoribbon produced a supercapacitor which offered 1241.5W/kg while maintaining a 26.9Wh/kg energy-density, with about 95% capacitance retention after 3000 cycles at 2.3A/g. A 1-pot hydrothermal technique[101] combined the growth of Fe_3O_4 nanoparticles with the reduction of graphene oxide, and the Fe_3O_4|reduced-graphene-oxide composite was then used as the electrodes of a supercapacitor which had a specific capacitance of 451F/g at a scan-rate of 5mV/s. The material had a high surface area, and an external magnetic field of 0.125T caused electrolyte ions to penetrate deeper and thus improve the capacitance. Electrodes within such a field had a specific capacitance of 868.89F/g; a 1.93 times resultant increase. The energy-density and power-density of the electrode in the presence of the magnetic field were 120.68Wh/kg and 3.91kW/kg, respectively.

Graphene nanosheets were decorated with Fe_3O_4 nanospheres by using a solvothermal method[102]. When used as an electrode, the nanocomposite had a specific capacitance of 268F/g at 2mV/s, with 98.9% capacitance retention after 10000 cycles. A resultant graphene|MnO_2 versus graphene|Fe_3O_4 asymmetrical supercapacitor offered an energy-density of 87.6Wh/kg, with 93.1% capacitance retention after 10000 cycles. A flexible nanocomposite was prepared[103] via the electrophoresis of a reduced graphene oxide network on carbon nanotube-Fe_3O_4 film. The interconnected network of graphene, with a specific surface area of 248.4m^2/g, improved the diffusion of electrolyte ions into the electrode. A resultant supercapacitor offered a specific energy-density of 36.7Wh/kg, a specific capacitance of 275.6F/g at a current density of 1A/g, and 92.8% capacitance retention after 10000 cycles.

A supercapacitor was constructed[104] by building a 3-dimensional reduced graphene oxide structure onto a carbon-nanotube|Fe_3O_4|polyaniline film. The electrodes had a specific capacitance of 414.5F/g at 1A/g. A resultant symmetrical supercapacitor which was made from the electrodes offered an energy-density of 60.8Wh/kg and a power-density of 45.2kW/kg. Nanoparticles of Fe_3O_4 were electrochemically deposited[105] onto N-doped porous graphene nanosheets. Electrodes of Fe_3O_4|Ni-foam and Fe_3O_4|N-doped-porous-graphene|Ni-foam had a specific capacitance as high as 822 and 631F/g at 0.5 and 10A/g, respectively. The pristine Fe_3O_4|Ni-foam electrode had values of 279 and 131F/g at 0.5 and 10A/g.

H_xWO_3

A composite of reduced graphene oxide decorated[106] with hydrogen tungsten oxide had an areal capacitance of 409mF/cm^2 at a 1mA/cm^2 current density. The composite was drop-cast onto a graphite layer on scotch tape to produce a flexible supercapacitor. The latter had a volumetric capacitance of 3.56F/cm^3 and offered an energy-density of 1.6mWh/cm^3 at a 1469.43mW/cm^3 power-density.

$KMnO_4$

A new method for synthesizing oxygen-containing porous graphene involved using $KMnO_4$ as an activator precursor[107]. Optimized samples possess a crumpled porous framework having a suitable surface area and many oxygen-containing functional groups. Electrodes had a gravimetric and volumetric capacitance of 363.3F/g and 342.5F/cm^3 at 0.5A/g, and 240F/g and 226.3F/cm^3 at 30A/g. A resultant symmetrical supercapacitor offered a volumetric energy-density of 19.2Wh/l.

$K_7MnV_{13}O_{38}$

Composites of the form, $K_7Mn^{IV}V_{13}O_{38} \bullet 18H_2O$|graphene-oxide, with ratios of 2:1 or 4:1, were used[108] as electrodes in supercapacitors. Composites with a 2:1 ratio in 1M LiCl electrolyte and with a 4:1 ratio in 1M Na_2SO_4 electrolyte had specific capacitances of 269.15F/g and 387.02F/g, respectively and energy-densities of 37.38Wh/kg and 53.75Wh/kg, respectively at a scan-rate of 5mV/s.

$La_2Ti_2O_7$

Graphene-oxide|$La_2Ti_2O_7$ fiber composites were prepared[109] by using a hydrothermal method. The titanate fibers were homogeneously dispersed between the graphene-oxide sheets. The composite had a specific capacitance of 900.6F/g at a current density of 1A/g in 1M H_2SO_4 with 10wt% sucrose aqueous solution as the electrolyte. With a potential window of 1.8V, an asymmetrical supercapacitor offered a maximum energy-density of 94.0Wh/kg at a power-density of 750.1W/kg.

$Li_4Ti_5O_{12}$

A supercapacitor was constructed[110] with oxide nanosheet as the negative electrode and N-doped graphene aerogel composite as the positive electrode. The device offered a maximum energy-density of 70Wh/kg with a power-density of 200W/kg, and an energy-density of 21Wh/kg at a power-density of 8kW/kg; with of 64% retention after 10000 cycles at a current density of 1.5A/g. Free-standing $Li_4Ti_5O_{12}$|graphene-foam composite[111] was used as the anode of a Li-ion hybrid supercapacitor. As-synthesized composite had specific capacitances of 186, 179 and 175mAh/g at 0.2, 0.5 and 1C, respectively. Supercapacitors which were constructed with the composite as the anode and activated-carbon as the cathode offered energy densities of 46 and 26Wh/kg at power densities of 625 and 2500W/kg, respectively; with 83% capacitance retention after 4000 cycles at 1A/g. Three-dimensional flower-shaped $Li_4Ti_5O_{12}$|graphene nanostructures[112] had a specific capacitance of 706.52F/g at 1A/g. An asymmetrical supercapacitor which had the composite as the positive electrode and carbon as the negative electrode offered an energy-density of 35.06Wh/kg at a power-density of 800.08W/kg; with 90.18% capacitance retention after 2000 cycles.

$MgAl_2O_4$

A $MgAl_2O_4$|reduced-graphene-oxide composite had a framework which impeded aggregation of the oxide and re-stacking of the graphene[113]. The composite had a specific capacity of 536.6F/g at 1A/g, and 257.3F/g at 40A/g; with 96.9% retention after 10000 cycles at 5A/g. An asymmetrical supercapacitor, having electrodes made of the composite and of activated-carbon offered an energy-density of 16.2Wh/kg at a power-density of 400kW/kg.

$MnCo_2O_4$

A flexible electrode of $MnCo_2O_4$|graphene was produced[114] by intercalating $MnCo_2O_4$, derived from Mn-doped ZIF-67 into graphene oxide sheets, followed by hydrothermal reduction. An asymmetrical supercapacitor was constructed which had activated-carbon and the composite as the negative electrode and positive electrode, respectively. The specific capacitance of the composite electrode could attain 1467F/g at a current density of 1A/g and offer an energy-density of 32.7Wh/kg at a power-density of 6250W/kg; with 83.9% capacitance retention after 9000 cycles.

$MnFe_2O_4$

A ternary $MnFe_2O_4$|graphene|polyaniline composite, in which the $MnFe_2O_4$ particles were dispersed on a flexible graphene sheet and wrapped in polyaniline, was used[115] as the negative electrode in a supercapacitor. The ternary composite had a specific capacitance of 241F/g at 0.5mA/cm². The supercapacitor had a maximum specific capacitance and energy-density of 48.5F/g at 0.5mA/cm² and 17Wh/kg, respectively.

MnO_2

Transition metal oxides with many reversible oxidation states are generally considered to be potential active materials for supercapacitor applications. Combining manganese dioxide, with its theoretical energy-density of 308Wh/kg, has great potential as supercapacitor electrodes for energy-storage applications, but a low capacitance use and capacitance fading arises from a poor electroconductivity and irreversible phase transformation. Sulfur-doping is a promising approach to the generation of electron-dense active regions on the surface of graphene-based materials. Graphene can also be decorated with flower-like MnO_2 nanostructures by means of electrodeposition. The nano-flowers consisted of tiny rods having a thickness of less than 10nm. When as-prepared graphene and MnO_2 were incorporated into an asymmetrical supercapacitor[116] the specific capacitance of the graphene electrode attained 245F/g at a charging current of 1mA, following electro-activation; which had made that magnitude 60%. The specific capacitance following MnO_2 deposition was 328F/g at a charging current of 1mA. The energy-density was 11.4Wh/kg and the power-density was 25.8kW/kg.

A thin layer of MnO_2 was deposited onto chemically-modified graphene[117]. The porous graphene structure with its large surface area aided rapid ionic transport within the electrode and led to a specific capacitance of 389F/g at 1A/g, with 97.7% capacitance retention upon increasing the current to 35A/g. When the MnO_2|graphene electrode was combined with a plain graphene electrode, the device offered an energy-density of 44Wh/kg and a power-density of 25kW/kg. An asymmetrical supercapacitor was constructed[118] by using graphene hydrogel with 3-dimensional interconnected pores as the negative electrode, and vertically-aligned MnO_2 nanoplates on nickel foam as the positive electrode, in a neutral aqueous Na_2SO_4 electrolyte. The device offered an energy-density of 23.2Wh/kg with a power-density of 1.0kW/kg. The energy-density of the asymmetrical supercapacitor was markedly greater than that of a symmetrical supercapacitor which was based upon graphene hydrogel (5.5Wh/kg) or MnO_2|Ni-foam (6.7Wh/kg). At a power-density of 10.0kW/kg, the asymmetrical device offered an energy-density of 14.9Wh/kg, with 83.4% capacitance retention after 5000 cycles.

Commercial graphene powder has been used[119] as a high surface-area substrate and capacitive electrode together with an aqueous electrolyte of 0.5M H_2SO_4 and 0.5M $MnSO_4$. A deposition-dissolution reaction of the MnO_2 provided most of the charge-storage in the battery-like electrode. Energy- and power-densities of 25Wh/kg and 980W/kg, respectively, were maintained for at least 5000 cycles. The electrolyte-volume to electrode-mass ratio was a critical factor governing the possible charge/discharge cycles, with the efficiency of the MnO_2 electro-dissolution/deposition process being less than 100%. An energy-density of 25Wh/kg was possible. Nanoparticles of MnO_2 were uniformly distributed and tightly anchored on graphene by using an ethanol-assisted reduction method[120]. This produced composite electrodes having an energy-density which could be varied up to 12.6Wh/kg at a power-density of up to 171kW/kg.

A nanocomposite of MnO_2, carbon nanotubes and graphene oxide was produced[121] by using graphene oxide as a surfactant to disperse pristine carbon nanotubes before the deposition of MnO_2 nanorods. This nanocomposite was used as a supercapacitor electrode which had a specific capacitance that was 4.7 times higher than that of free MnO_2. The energy-density was 30.4 to 14.2Wh/kg and the power-density was 2.6 to 50.5kW/kg; with some 94% capacitance retention after 1000 cycles. A supercapacitor electrode material composed of MnO_2 and reduced graphene oxide, coated onto flexible carbon-fiber paper, was produced[122] by colloidally mixing reduced graphene oxide nanosheets and 1.8nm MnO_2 nanoparticles. The specific capacitance of the electrode material at 0.1A/g was about 393F/g, with 98.5% capacitance retention after 2000 cycles. Composites of MnO_2 nanorods with graphene have been fabricated[123] by using a hydrothermal method. The maximum specific capacitance was 218F/g at a scan-rate of 5mV/s in 1M aqueous Na_2SO_4 solution. The composite exhibited a maximum energy-density of 16Wh/kg at a power-density of 95W/kg; with some 94% capacitance retention after 1000 cycles.

Asymmetrical supercapacitor cells have been based[124] upon the use of MnO_2 as the positive electrode and graphene as the negative electrode in aqueous 1M Na_2SO_4 electrolyte. Amorphous MnO_2 nanoparticles were prepared by reducing $KMnO_4$ in N,N-dimethylformamide. The graphene was prepared by the hydrothermal reduction of exfoliated graphene oxide sheets. The MnO_2/graphene asymmetrical supercapacitor could operate reversibly at a cell voltage of 2.0V, with an energy-density of 25.2Wh/kg and a power-density of 100W/kg; with 96% capacitance retention after 500 cycles.

So-called dip-and-dry methods have been used[125] to prepare graphene/MnO_2 nanostructured sponges for use as supercapacitor electrodes. Commercial sponges were used as skeletons for the construction of homogeneous 3-dimensional interconnected macro-network reduced graphene oxide composites. These could operate at a scan-rate of 200V/s and exhibited about 10% degradation after 10000 cycles at a charge-discharge specific current of 10A/g. Sponge-based supercapacitors retained some 90% of their capacitance after 10000 cycles at a scan-rate of 10V/s. The maximum energy-density and highest power-density of a reduced graphene oxide sponge-based device were 2.08Wh/kg and 94kW/kg, respectively, at an operating voltage of 0.8V. The corresponding values were 8.34Wh/kg and 47kW/kg for sponge-based reduced graphene oxide plus MnO_2 composite at an operating voltage of 0.8V.

Composites of $Ni(OH)_2$, MnO_2 and reduced graphene oxide have been used[126] as supercapacitor electrodes. They had a specific capacitance of 1985F/g, and an energy-density of 54.0Wh/kg. Composites of MnO_2, reduced graphene oxide, indium tin oxide and polypyrrole were prepared[127] by the chronopotentiometric deposition of manganese dioxide or polypyrrole onto reduced-graphene-oxide|indium-tin-oxide film. Asymmetrical supercapacitors which were constructed by using the two composites, as positive and negative electrodes in polyvinylalcohol-LiCl gel electrolyte, offered a power-density of 7.4kW/kg at an energy-density of 13Wh/kg and an energy-density of 16Wh/kg at a power-density of 0.3kW/kg; with 75% capacitance retention after 2000 cycles.

Reduced graphene oxide, added to polyvinylidene fluoride films, has been used[128] as a substrate for the successive electrodeposition of MnO_2 and carbon nanotube layers so as to form flexible supercapacitor electrodes. These films maintained the high conductivity of reduced graphene oxide nanosheets, the high surface:volume ratio of carbon nanotubes and the large pseudocapacitance of MnO_2 nanosheets. Asymmetrical supercapacitors which were made from the present film electrode, a platinum sheet electrode, and 6M NaOH electrolyte had a working voltage of up to 1.8V together with an energy-density of 49.1Wh/kg at a power-density of 4.83kW/kg. The specific capacitance was 276.3F/g at a discharge current density of 0.08A/g. Graphene bubbles were synthesized[129] by chemical vapor deposition and metal (Fe, Cu, Fe/Cu) supported mesoporous catalyst. The growth of bubbles was achieved by using isobutane gas as the carbon source at 900C. The metal-doped catalyst had a surface area of $1380m^2/g$, with a pore-diameter of 6nm. The resultant graphene structure had a specific surface area of $400m^2/g$, with a bubble morphology, and MnO_2 was deposited onto the graphene surface. Electrodes which were prepared by using Cu, Fe and Fe/Cu supports had specific capacitances of 228, 257 and 327F/g, respectively, with energy-densities of 18, 16 and 22Wh/kg at a scan-rate of 100mV/s.

A 3-dimensional ternary composite, graphene|polyaniline|MnO_x, had a specific capacitance of 955F/g at a current density of 1A/g, with 89% capacitance retention after 1000 cycles and 69.1% retention after 5000 cycles at $20A/g^{130}$. Electrodes which were prepared from this material offered an energy-density of 61.2Wh/kg at a power-density of 4.5kW/kg. Reduced graphene oxide, MnO_2 and carbon black have similarly been used[131] to prepare film via vacuum filtration. The MnO_2 nanoparticles grew on both sides of the graphene layers and served as active sites for electrochemical reactions which markedly increased the specific capacitance. The carbon black spaced the graphene layers apart and prevented re-stacking; thus improving the conductivity between the basal planes of the graphene sheets. Because of the synergistic effects of carbon black and MnO_2, the capacitance was increased to 209F/g, as compared with the 96F/g of binary graphene/carbon-black film, and was maintained at 77F/g at a scan-rate of 1V/s. A 1.8V aqueous asymmetrical supercapacitor which was based upon the new film as the positive electrode and reduced graphene oxide and carbon black as the negative electrode exhibited an energy-density of 24.3Wh/kg at high power-delivery and an energy-density of 10Wh/kg at a power-density of 45kW/kg. The energy-density of this flexible solid-state asymmetrical device attained 20Wh/kg, and remained at 10Wh/kg for a power-density of 21kW/kg.

Layered graphene/MnO_2 nanocomposites have been prepared[132] by using a face-to-face electrostatic self-assembly method. The high-quality graphene was obtained via the 1-step electrochemical exfoliation of graphite. The face-to-face contact could improve the conductivity of the MnO_2 and also prevent the re-stacking and agglomeration of graphene. The composite had a specific capacitance of 319F/g at 0.2A/g, with 95.6% capacitance retention after 10000 cycles. The composite offered an energy-density of 31.5Wh/kg at a power-density of 100W/kg. Graphene oxide nanosheets have been simultaneously reduced, and deposited onto nickel foam, by using a 1-step electrodeposition method[133]. The resultant interconnected and crumpled graphene nanosheets then served as a 3-dimensional conductive skeleton for the hydrothermal deposition of MnO_2 nanosheets. The MnO_2 nanosheets which were anchored to the graphene-covered nickel foam formed an unique 3-dimensional porous interconnected network. The Ni-foam/graphene/MnO_2 electrodes had a specific capacitance of 462F/g at 0.5A/g with 93.1% retention after 5000 cycles at 10A/g, together with an energy-density of 26.1Wh/kg at a power-density of 3981W/kg.

Three-dimensional porous graphene plus MnO_2 composites were prepared[134] by the deposition of MnO_2 particles onto graphene produced by freeze-drying and used as electrodes for supercapacitors. Upon immersing the porous graphene in 0.1M $KMnO_4$/K_2SO_4 for various times, MnO_2 particles having a size of about 200nm were uniformly on the grapheme sheets. Composite which was immersed for 2h had the highest specific capacitance of 800F/g, with the highest energy-density of 40Wh/kg at a current density of 0.1A/g. Composites of MnO_2 and electrochemically-reduced graphene oxide have been made[135] by electrophoretic deposition from an aqueous dispersion of graphene onto stainless steel. Dioxide nanoparticles with dimensions of 40 to 70nm were formed a uniform coating. The specific capacitance of MnO_2/graphite/steel composites was 392F/g at a current density of 1A/g in 0.5M Na_2SO_4, as compared with the 159F/g in the absence of the dioxide. The power- and energy-densities were also higher.

Highly-aligned MnO_2 nanowall arrays have been electrodeposited onto titanium sheets and used as substrates for the growth of graphene by chemical vapor deposition and the formation of a 3-dimensional oxide/graphene hybrid composite[136]. The as-prepared material could be directly used as a supercapacitor electrode without requiring any binder. In an aqueous electrolyte the hybrids had a specific capacitance of about 326.33F/g at a scan-rate of 200mV/s, together with an energy-density of 23.68Wh/kg at a power-density of 7270W/kg. Free-standing 2-dimensional MnO_2/carbon-sphere/graphene film has been used[137] as an electrode material for a supercapacitor. The carbon spheres spaced sandwich-like porous structures and provided paths for electrolyte-ion penetration. Following decoration with MnO_2, the films could have a gravimetric capacitance of 319.3F/g and a volumetric capacitance of 277.8F/cm^3. The electrodes also had gravimetric and volumetric energy-densities of 29.4Wh/kg and 25.6Wh/l, respectively, with 94.1% retention after 5000 charge-discharge cycles at a current density of 5A/g.

Following electrochemical coating[138] with MnO_2, graphene-pellet|MnO_2 electrodes had specific and volumetric capacitances of up to 395F/g and 230F/cm^3, respectively, at 1A/g. When combined with hydroquinone and benzoquinone additive electrolytes, the material had a specific capacitance of 7813F/g at 10A/g. When the graphene-pellet|MnO_2 electrode was combined with a graphene-pellet|polypyrrole electrode, the device offered a maximum energy-density of 26.7Wh/kg and a maximum power-density of 32.7kW/kg. Composites made of MnO_2, decorated onto porous carbon nanofiber/graphene, have been prepared by electrospinning[139]. The presence of graphene in the composite fibers aided the uniform dispersion of oxide particles and prevented their agglomeration. A graphene concentration of 5wt% offers led to a larger accessible specific surface area and good conductivity. A supercapacitor electrode which contained 5wt% graphene exhibited a specific capacitance of 210F/g at a current density of 1mA/cm^2, with 170F/g at a current density of 20mA/cm^2; combined with an energy-density of 24 to 19Wh/kg at power-densities ranging from 400 to 10000W/kg in 6M KOH aqueous solution.

Composites of MnO_2/polyaniline/graphene for use as supercapacitor electrodes have been synthesized by interfacial polymerization at an oil/water interface[140]. The 3-dimensional nanostructure featured a loose nanorod arrangement of polyaniline, coated with round MnO_2 pellets. The maximum energy-density was 45.4Wh/kg at a power-density of 67.8kW/kg, and the highest power-density was 229.2kW/kg at an energy-density of 25.7Wh/kg; with 70.4% and 59.1% retention after 500 cycles at a scan-rate of 5mV/s for oxide/graphene and polyaniline/graphene composites, respectively. Free-standing polyaniline/graphene/MnO_2 paper was constructed[141] by using low-cost printing paper, layer-by-layer *in situ* growth and vacuum filtration. The polyaniline/graphene paper surface was scattered with MnO_2 nanoflowers. The paper had a mass-loading of 11.75mg/cm^2, was highly flexible and could serve as an electrode without the using of a binder. The electrode had an areal capacitance of 3.5F/cm^2 at 5mA/cm^2, with 90% retention after 1000 bending cycles. A solid-state supercapacitor which was constructed by using the paper offered an energy-density of 5.2mWh/cm^3 at 8.4mW/cm^3.

Graphene/carbon-nanofiber/MnO_2 nanocomposite paper has been prepared[142] by chemical reaction between $KMnO_4$ and carbon in acidic solution, so that MnO_2 nanosheets were uniformly deposited onto the nanofiber paper. The material could be used as an electrode without adding a binder. The specific capacitance was 298.2F/g at 1A/g in 1M Na_2SO_4 electrolyte. The paper composite could handle high scan-rate loads with no large decrease in capacitance, with the specific capacitance maintained at 370.4F/g at 300mV/s. There was 95.7% capacitance retention after 3000 cycles. An asymmetrical supercapacitor with an operating voltage of 2.0V exhibited an energy-density of 23.0Wh/kg at 450.8W/kg. A nitrogen-doped graphene/MnO_2 nanosheet composite was prepared[143] by using a simple hydrothermal method. The N-doped graphene acted as a template for the growth of layered δ-oxide. The composite had a specific capacitance of about 305F/g at a scan-rate of 5mV/s. Work in this field can often be rather 'bricolagic': in this case, a cheap and highly conductive flexible current collector was made by using Scotch tape. Other researchers have reported[144] the preparation of reduced graphene oxide paper by using a household non-stick frying pan. Another preparation-method was said to have been inspired by bread-making[145]. Here, flexible solid-state asymmetrical supercapacitors were prepared from the present electrode material and a polyvinylalcohol-LiCl gel electrolyte. These devices could offer an operating voltage of 1.8V and a maximum energy-density of 3.5mWh/cm^3 at a power-density of 0.019W/cm^3; with better than 90% capacitance retention after 1500 cycles.

Three-dimensional MnO_2/graphene/carbon-nanotube hybrids have been obtained[146] by combining electrochemical deposition, electrophoretic deposition and chemical vapor deposition techniques. This led to a specific capacitance of 330.75F/g and an energy-density of 36.68Wh/kg together with a power-density of 8000W/kg at a scan-rate of 200mV/s. The specific capacitance was 187.53F/g at a scan-rate of 400mV/s. Similar studies[147], performed using nickel foam as a substrate led to a specific capacitance of up to 377.1F/g at a scan-rate of 200mV/s, with an energy-density of 75.4Wh/kg and 90% capacitance retention after 500 cycles. An asymmetrical supercapacitor was prepared[148] which featured flower-like MoS_2 and MnO_2 that were grown onto graphene nanosheets to furnish negative and positive electrodes, respectively. The MoS_2/graphene electrode exhibited a specific capacitance of 320F/g at 2A/g. The all-solid-state capacitor offered an energy-density of 78.9Wh/kg at a power-density of 284.1W/kg.

A 2-dimensional free-standing flexible MnO_2/graphene film supercapacitor electrode has been prepared[149] by using spin-coating and hydrothermal techniques. Oxide nanosheets were aligned vertically on one side of the graphene thin film. The graphene film acted as a substrate upon which MnO_2 nanosheets grew *in situ*. The material exhibited a good performance in a 3-electrode configuration, including a specific capacitance of up to 280F/g with no apparent decay after 10000 cycles. A symmetrical supercapacitor had a specific capacitance of up to 77F/g at a cell voltage of 1.0V, with 91% retention after 10000 cycles at a current density of 1A/g. It offered an energy-density of 10.7Wh/kg at a power-density of 500W/kg. Functionalized graphene is a good substrate for anchoring MnO_2 nanoflowers by electrodeposition to give an optimum weight-ratio of functionalized graphene and oxide of 1:8. The resultant electrodes have a specific capacitance of 320.59F/g at a current density of 0.5A/g, with 95.5% capacitance retention after 3000 cycles. Symmetrical supercapacitors which comprised the hybrid materials[150] exhibited a specific capacitance of 55.37F/g at a scan-rate of 5mV/s, together with a maximum energy-density of 5.67Wh/kg and a power-density of 5.11kW/kg. Nanoparticles of MnO_2 were electrodeposited onto graphene/activated-carbon composite film, with the urchin-type microspheres being controlled by adjusting the electrodeposition reaction-time[151]. The composite electrodes exhibited a maximum specific capacitance of 1231mF/cm^2 for an oxide loading of 7.65mg/cm^2, and a mass specific capacitance of 123F/g at a current density of 0.5mA/cm^2. A flexible solid-state symmetrical supercapacitor retained 88.6% of its original capacitance after 500 bending times, and retained some 82.8% of its capacitance after 10000 cycles. The device offered a maximum energy-density of $0.27mWh/cm^3$ and a maximum power-density of $0.02W/cm^3$.

A composite which comprised MnO_2 nanoparticles and 3-dimensional N-doped reduced graphene oxide was prepared by using hydrothermal and ultrasound techniques[152]. When used as a supercapacitor electrode material, the specific capacitance was 349F/g at 0.5A/g in a 1.0M Na_2SO_4 electrolyte solution; with the capacitance being retained after 20000 cycles. An asymmetrical supercapacitor made using MnO_2/graphene-oxide and graphene oxide as the positive and negative electrodes, respectively, exhibited an energy-density of 35.22Wh/kg at 200W/kg. An asymmetrical supercapacitor was constructed[153] by using porous manganese oxide nanostructures as the positive electrode and reduced graphene oxide as the negative electrode in 1M Na_2SO_4 electrolyte. The device offered an energy-density of 22.2Wh/kg at a power-density of 101W/kg, when cycled reversibly within the potential range of 0 to 2V; with 90% capacitance retention after 3000 cycles.

An asymmetrical supercapacitor has been constructed[154] by using reduced graphene oxide plus polymer and reduced graphene plus MnO_2 nanocomposites as the negative and positive electrodes, respectively. Cyclic voltammetry and galvanostatic charge-discharge data indicated a maximum specific capacitance of 247F/g at 1A/g, between −0.9 and $0.1V_{SCE}$. The graphene/oxide nanocomposite electrode had a specific capacitance of 145F/g at 1A/g, between −0.1 and $0.9V_{SCE}$. The asymmetrical supercapacitor had a cell voltage of 1.8V in 1M Na_2SO_4 electrolyte. The device capacitance was 47F/g, with an energy-density of 21Wh/kg.

A ternary polymer composite has been constructed[155] by combining graphene platelets, poly(3,4-ethylenedioxythiophene) and MnO_2 into a kitchen sponge (another example of the above-mentioned bricolage). Graphene platelets with an electrical conductivity of 1460S/cm were deposited onto the sponge surface by simple dipping and drying, followed by *in situ* polymerization of the 3,4-ethylenedioxythiophene. The dioxide was then deposited onto the prior layers by soaking the sponge in $KMnO_4$ solution. The composite was used as electrodes in a supercapacitor having a stretchability of up to 400%. The supercapacitor had a specific capacitance of 802.99F/g, together with an energy-density of 55.76Wh/kg and 99% capacitance retention after 1000 stretching cycles. The mass and thickness of the oxide layer increased with soaking time, yielding values of 368.96F/g for 300s and 740.25F/g for 600s. A commercial kitchen sponge was also used[156] as a scaffold upon which graphene platelets and polyaniline nanorods were deposited. The high electrical conductivity of the platelets enhanced the pseudo-capacitive performance of polyaniline grown vertically on the graphene basal planes, while the interconnected pores of the sponge facilitated ion diffusion. The composite electrode was used to construct a supercapacitor which had a specific capacitance of 965.3F/g at a scan-rate of 10mV/s in 1M H_2SO_4 solution. At 100mV/s, the supercapacitor offered an energy-density of 34.5Wh/kg and a power-density of 12.4kW/kg.

A flexible asymmetrical supercapacitor was constructed[157] by using a composite of hollow urchin-shaped coaxial MnO_2 and polyaniline as the positive electrode and 3-dimensional graphene foam as the negative electrode in a polyvinylalcohol/KOH gel electrolyte. Dioxide/polyaniline and graphene-foam/graphene-foam symmetrical supercapacitors exhibited specific capacitances of 129.2 and 82.1F/g, respectively, at a current density of 0.5A/g. The asymmetrical supercapacitor offered a higher energy-density (37Wh/kg) than did the symmetrical supercapacitors, where the energy-density was 17.9 and 11.4Wh/kg for dioxide/polyaniline/dioxide/polyaniline and graphene-foam/graphene-foam, respectively, at 0.5A/g. The asymmetrical supercapacitor exhibited an energy-density of 22.3Wh/kg at 5A/g, with 89% capacitance retention after 5000 cycles. Three-dimensional MnO_2/reduced-graphene-oxide aerogels have been prepared[158] by using hydrothermal methods, and used as the electrode of a supercapacitor. Asymmetrical devices were constructed with aqueous Na_2SO_4 solution as the electrolyte, oxide/reduced-graphene-oxide aerogel as the positive electrode and reduced graphene oxide as the negative electrode. It offered a maximum energy-density of 18.2Wh/kg and a power-density of 400W/kg.

Hybrid fibers of transition-metal oxide nanorods and reduced graphene oxide were prepared[159] by wet-spinning method and an asymmetrical supercapacitor was constructed by using MnO_2-nanorod/reduced-graphene-oxide fiber as the positive electrode, MoO_3-nanorod/reduced-graphene-oxide fiber as the negative electrode and H_3PO_4/polyvinylalcohol as the electrolyte. The asymmetrical supercapacitor could be cycled reversibly at a voltage of 1.6V and deliver a volumetric energy-density of 18.2mWh/cm^3 at a power-density of 76.4mW/cm^3. An ionic layer adsorption and reaction technique can be used for the layer-by-layer assembly of reduced graphene oxide and MnO_2 on a stainless-steel current collector for supercapacitor electrodes. An asymmetrical supercapacitor[160] which had MnO_2/reduced-graphene oxide as the positive electrode and thermally-reduced graphene oxide as the negative electrode offered an energy-density of about 88Wh/kg, a power-density of about 23200W/kg and some 79% capacitance retention after 10000 charge-discharge cycles.

Films of MnO_2 and reduced graphene oxide which were prepared[161] by filtration deposition and thermal reduction had a specific capacitance of 333.9F/g at 0.5A/g, with 87% retention after 3000 cycles at 0.5A/g. Flexible symmetrical supercapacitors were constructed by using these films as electrodes. The device had a safe working range of 0 to 0.7V, with a maximum energy-density of 23.5Wh/kg at 0.5A/g and a maximum power-density of 1716.9W/kg at 2.25A/g.

There is an advantage to adding more than one transition-metal oxide because the simultaneous presence of 3d and 4d groups expectedly improves the properties of a composite[162]. The addition of MnO_2-RuO_2 nanoflakes to reduced graphene oxide nanoribbon formed a network having enhanced diffusion kinetics and thus better supercapacitor performance. An asymmetrical device offered an energy-density of 60Wh/kg at a power-density of 14kW/kg.

A 3-dimensional graphene hydrogel was prepared[163] by using a self-assembling process in glucose and ammonia, followed by freeze-drying. A δ-MnO_2/hydrogel composite was then obtained by immersing the hydrogel in $KMnO_4$ solution. An asymmetrical supercapacitor of the form, MnO_2/hydrogel/hydrogel, consisted of pseudocapacitive nanosheet-like δ-MnO_2/graphene-hydrogel as the cathode and electric double-layer capacitive graphene hydrogel as the anode. This offered an energy-density of 34.7Wh/kg at a power-density of 1.0kW/kg. A composite electrode was prepared[164] by electrodepositing MnO_2 into a graphene hydrogel, and this had a specific capacitance of 352.9F/g at 1A/g. An aqueous asymmetrical supercapacitor, which comprised activated-carbon as the negative electrode and MnO_2|graphene-hydrogel as the positive electrode, exhibited 91.5% capacitance retention after 5000 cycles. It offered an energy-density of 41.8Wh/kg at a power-density of 0.2kW/kg.

A nanocomposite of Mn_3O_4-nanocrystals plus reduced graphene oxide was prepared by using a hydrothermal method[165]. There was formation of the tetragonal hausmannite phase of Mn_3O_4 nanocrystals and of reduced graphene oxide sheet/scrolls. The material had a specific capacitance of 611F/g, with 95% capacitance retention after 3000 cycles. Polypyrrole films were deposited onto flexible carbon fiber by electrodeposition. An asymmetrical supercapacitor, with the oxide plus graphene composite as the positive electrode and polypyrrole as the negative electrode, offered a maximum energy-density and power-density of 32Wh/kg and 833W/kg, respectively, with 90% capacitance retention after 6000 cycles.

Porous carbon nanosheets were prepared[166] via 1-step activation and carbonization of naturally hollow tube-like dandelion-down, the hollow tube structure of which was composed of aligned nanocellulose. The dandelion down was converted into porous interconnected carbon nanosheets. Porous carbon nanotubes combined with MnO_2 were used as the positive electrode in an asymmetrical supercapacitor. The latter offered an energy-density as high as 28.2Wh/kg at a power-density of 899.36W/kg, with 89%capacitance retention after 10000 cycles.

A nanocomposite was synthesized[167] which combined poly(3,4-ethylene dioxythiophene polystyrene sulfonate, MnO_2 nanowires and graphene oxide. The maximum specific areal capacitance was $841F/g$ ($177mF/cm^2$) at $10mV/s$, with an energy-density of $0.593kWh/kg$ at $0.5mA$. Ternary composites of MnO_2, activated carbon and graphene have been used[168] as supercapacitor electrodes. Their preparation involved the chemical vapor deposition of graphene onto nickel foam, followed by decorating activated carbon onto the graphene surface by dip-coating. A porous MnO_2 upper layer on the ternary composite was formed via self-limiting growth involving redox reactions of $KMnO_4$ on the activated-carbon layer. A composite electrode with $20mM$ $KMnO_4$ exhibited a maximum specific capacitance of $813.0F/g$ at a current density of $1.0A/g$, with 98.4% capacitance retention after 1000 cycles. A solid state supercapacitor which was based upon the ternary composite electrodes offered an energy-density of $33.9Wh/kg$ and a power-density of $319.3W/kg$.

An asymmetrical supercapacitor was prepared[169] by using graphene|MnO_2 nanosheet as the cathode and porous graphene as the anode. The MnO_2 ultra-thin nanosheets were formed on the graphene sheets by polyaniline-assisted growth. The graphene|MnO_2 electrode had a specific capacitance of up to $245.0F/g$ at $0.5A/g$, with 74.5% retention ratio at $20A/g$. The resultant asymmetrical supercapacitor could offer an energy-density of $7.9Wh/kg$ at a power-density of $11804W/kg$; with 91.5% capacitance retention after 10000 cycles. Phytic acid linked MnO_2/reduced-graphene-oxide composite has been prepared[170] by using a self-assembling hydrothermal and freeze-dry method. The acid accelerated the immobilization of MnO_2 on the graphene-oxide sheets to form a composite aerogel having a 3-dimensional layered porous structure. When the composite aerogel was used as supercapacitor electrodes, the device could attain a specific capacitance of $645F/g$ with a current-density of $1A/g$, and $410F/g$ at $40A/g$; with 94.9% capacitance retention after 10000 cycles under a current density of $20A/g$. An asymmetrical supercapacitor offered a maximum energy-density of $59.2Wh/kg$ at $1.48kW/kg$, remaining at $28.8Wh/kg$ for a maximum power-density of $57.6kW/kg$; with 91.6% retention after 10000 cycles at $20A/g$.

A ternary nanocomposite of the form, MnO_2|reduced-graphene-oxide|quantum-dot polyaniline, exhibited[171] a specific capacitance of $423F/g$ at a current density of $5.7A/g$ in a 3-electrode configuration. The specific capacitance of the nanocomposite electrode was much higher than that of the individual components. The capacitance retention was almost 85% after 2000 charge–discharge cycles. An asymmetrical device was constructed by using the nanocomposite as the positive electrode and graphene oxide as the negative electrode. It offered a high energy-density without any great decrease in the power-density, with 90% capacitance retention after 2000 cycles. A 1-pot method was used[172] to produce highly exfoliated reduced graphene oxide plus manganese oxide nanocomposites. The manganese oxide nanoparticles were spherical in reduced graphene oxide plus Mn_3O_4 composites, and spherical or cubic in reduced graphene plus MnO_x composites. Highly-exfoliated reduced graphene plus MnO_x nanocomposite exhibited a higher electrochemical capacitive responsive than that of reduced graphene oxide plus Mn_3O_4 nanocomposite. The reduced graphene oxide plus MnO_x nanocomposite exhibited the highest capacitance of 398.8F/g at a sweep-rate of 5mV/s, with an energy-density of 23.3Wh/kg and a power-density of 2001W/kg.

In order to improve the capacitance of graphene paper, MnO_2 was electrochemically deposited[173] in the form of nanoflowers which then aided the fast transfer of electrolyte ions. Following 10 cycles of electrodeposition, the MnO_2-coated graphene paper exhibited a specific capacitance of 385.2F/g at 1mV/s in 0.1M Na_2SO_4, with high capacitance retention after 5000 cycles. A flexible solid-state asymmetrical supercapacitor was constructed by using graphene-paper|MnO_2 as the positive electrode and graphene-paper alone as the negative electrode. The device exhibited an areal capacitance of 76.8mF/cm^2 at a current density of 0.05mA/cm^2, with 82.2% retention after 5000 galvanostatic charge/discharge cycles. The flexible device also delivered an energy-density of 6.14mWh/cm^2 with a power-density of 36mW/cm^2. Nanocomposites with δ-MnO_2 nanoplatelets on N-doped reduced graphene oxide have been synthesized[174] by using hydrothermal methods, with p-phenylenediamine serving to anchor the δ-MnO_2 to the graphene oxide as well as acting as a nitrogen source. The composite had a specific capacitance of 299.5F/g at 5mV/s, with 97.8% retention after 8000 cycles. The energy-density was 24.2Wh/kg for a power-density of 242W/kg.

A composite of graphene with MnO_2 nanorods was obtained[175] via the hydrothermal oxidation of a manganese precursor on the graphene surface. The performance of the composite electrode in a symmetrical device revealed an energy-density of 42.7Wh/kg and a specific capacitance of 759F/g. The excellent performance was attributed to the composite structure, which provided passages for electrons but also increased ion transportation during fast charge-discharge reactions. A MnO_2/laser-induced-graphene electrode exhibited[176] a capacitance of 48.9mF/cm^2; some 2.4 times greater than that of a pure laser-induced graphene electrode. An all-solid supercapacitor which contained the present electrodes offered an energy-density of 3.1µWh/cm^2 and a power-density of 2.5mW/cm^2; with 94.3% retention after 3000 cycles at a current density of 0.5mA/cm^2. This performance was hardly affected by bending the device through 150°.

The MnO_2-based supercapacitors can suffer from problems such as its low electrical conductivity, its incomplete utilization and its dissolution. Nanoscroll structures of MnO_2/reduced-graphene-oxide hybrid have been suggested to overcome these difficulties. A supercapacitor which had MnO_2/ reduced-graphene-oxide nanoscrolls as the anode and activated carbon as the cathode exhibited[177] an improved energy-density and a lithium storage ability that bridged the energy–density gap between conventional Li-ion batteries and supercapacitors. The supercapacitor exhibited a lithium discharge capacity of 2040mAh/g, and a maximum energy-density of 105.3Wh/kg with a power-density of 308.1W/kg. It delivered an energy-density of 42.77Wh/kg at a power-density of up to 30800W/kg.

Manganese dioxide nanoflowers were obtained by recycling spent batteries, and single electrodes of MnO_2 nanoflowers and reduced graphene oxide nanosheets exhibited a specific capacitance of 208.5F/g and 145F/g, respectively. An asymmetrical supercapacitor which was made[178] from these materials and Na_2SO_4 had a specific capacitance of 177.6F/g and an energy-density of 24.7Wh/kg; with 95.2% retention after 4000 cycles. An asymmetrical supercapacitor was constructed[179] which had a negative electrode of graphene foam loaded polypyrrole nanowire and a positive electrode of polypyrrole|MnO_2 core–shell nanowires on graphene foam. It could be repeatedly discharged/charged at 1.6V, and offered an energy-density of $1.04mWh/cm^3$ together with good cycling stability. Manganese dioxide and MnO_2/reduced-graphene-oxide composite electrodes were prepared[180] without a binder under argon. The specific capacitance was 175F/g at a current density of 20A/g and the energy-density could attain 41.27Wh/kg at 1A/g.

Free-standing flexible electrodes were made[181] from MnO_2|polyaniline|carbon-nanotube|B-doped-graphene for supercapacitor use. The material had a specific capacitance of 1544F/g at 1A/g, with 84% retention after 5000 cycles. It also exhibited a power-density of 2353W/kg and an energy-density of 196Wh/kg at 1A/g. Nanorod/fiber, nanosheet and nano-flower structures could be synthesized[182] from S-reduced graphene oxide and MnO_2. Electrochemical measurements revealed a specific capacitance of 180.4F/g at 1A/g in 2.5M KNO_3 electrolyte. An asymmetrical device which had S-treated reduced graphene plus MnO_2 as the positive electrode, and activated carbon derived from peanut-shells as the negative electrode, offered a specific energy of 71.74Wh/kg and a corresponding specific power of 850W/kg at 1A/g. At a specific current of 5A/g, the device maintained a specific energy of 55.30Wh/kg. A capacitance retention of 94.5% was found after 10000 cycles at 5A/g.

Three-dimensional S-doped graphene has been prepared[183] by heat-treatment because uniformly-doped S provides more anchoring-points for mixed-valency manganese oxides. A S-doped graphene plus MnO_x composite was obtained via a 1-pot hydrothermal method. The specific capacitance of a composite electrode at 0.5A/g was 1311F/g, with 81.2% retention after 5000 cycles. An asymmetrical supercapacitor with composite and activated-carbon electrodes offered an energy-density of 21.23Wh/kg and a power-density of 1.5kW/kg. The biomass material, diatomite, has been used as a porous template for the *in situ* growth of graphene layers[184]. Carbon nanotube arrays were then grown on the graphene surface. Composites of MnO_2, carbon nanotubes, graphene and diatomite exhibited a maximum specific capacitance of 264.0F/g. An asymmetrical supercapacitor was constructed from MnO_2|nanotube|graphene|diatomite and microwave-exfoliated graphene-oxide electrodes offered a maximum energy-density of 64.3Wh/kg and a maximum power-density of 19.8kW/kg. Graphene-wrapped MnO_2 was prepared[185] by using a solution-based ultrasonic-assisted method. Porous oxide microspheres, wrapped with graphene nanosheet, had a specific surface area which aided rapid ion-diffusion and transport. The material had a capacitance of 1227F/g at 0.5A/g, with 90% retention after 5000 cycles at 2A/g. An asymmetrical supercapacitor which was based upon graphene-wrapped MnO_2 offered an energy-density of 19.6Wh/kg at a power-density of 351W/kg.

In situ growth of a 3-dimensional MnO_2 fiber network over the surface of graphene layers was achieved[186] at a solid/liquid interface. The specific capacitance attained 683F/g at a current density of 2.2A/g, with 96.9% capacitance retention after 5000 cycles. At a current density of about 10.86A/g, a power-density of about 6.235kW/kg was possible while maintaining an energy-density of about 7.99kWh/kg. Lanthanum-doped (1 to 5vol%) MnO_2 plus graphene oxide composite electrode thin films were prepared[187] via binder-free ionic-layer adsorption and reaction. A composite with 3%La-doped MnO_2 exhibited porous spongy nanoparticles, and mesoporous sheets had a surface area of up to 149m^2/g. The highest specific capacitance was 729F/g for this composition, at a scan-rate of 5mV/s; with 94% retention after 5000 cycles. A flexible symmetrical supercapacitor of the form,

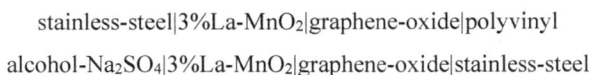

stainless-steel|3%La-MnO_2|graphene-oxide|polyvinyl

alcohol-Na_2SO_4|3%La-MnO_2|graphene-oxide|stainless-steel

operating with a potential window of 1.8V had a maximum specific capacitance of 140F/g, with an energy-density of 64Wh/kg at a power-density of 1kW/kg; with 90% capacitance retention after 5000 cycles at a scan-rate of 100mV/s. Free-standing MnO_2 on N-doped graphene and single-wall carbon nanotubes was produced[188] by electrochemical deposition. An as-prepared oxide|graphene cathode had a capacitance of 489.7F/g at 1A/g. A symmetrical aqueous supercapacitor had a voltage of 2.4V and an energy-density of 106.7Wh/kg at 1200W/kg; with no decay after 10000 cycles.

Mesoporous CeO_2|α-MnO_2|reduced-graphene-oxide composite was prepared[189] by hydrothermal means and used as an electrode for a supercapacitor. The specific capacitance of the optimum composition could attain 466F/g at a current density of 1.0A/g in 1M Na_2SO_4 solution, while the energy-density and power-density were 259Wh/kg and 1008W/kg, respectively; with essentially 100% retention after 10000 cycles. A composite of α-MnO_2 nanowires and N-doped reduced graphene oxide was used for the electrodes of a symmetrical superconductor[190]. Hydrothermally prepared α-MnO_2 nanowires had a smooth surface and a high length-to-width ratio. The composite film had electrode had a areal specific capacitance of 386.8mF/cm^2 at a 1mA/cm^2 current density; with 62.4% capacitance retention when the current density was increased from 1 to 10mA/cm^2. A flexible symmetrical capacitor with α-MnO_2/graphene-oxide film electrodes and a 1M Na_2SO_4 electrolyte with a cellulose paper separator, exhibited 94% capacitance retention after 3000 cycles, together with an energy-density of 28.22μWh/cm^2 at a power-density of 0.8mW/cm^2.

A hydrothermal method could be used to create α-MnO_2-nanotube|δ-MnO_2-nanoflake structures on graphene foam[191]. When the MnO_2 content was 60wt%, the maximum specific capacitance was 202F/g with respect to the entire electrode and 336F/g relative to active material; with 97% capacitance retention after 5000 cycles. An asymmetrical supercapacitor which was based upon the foam|nanotube|nanoflake composites offered an energy-density of 23.2W/kg under a power-density of 119.9W/kg, over a potential range of 0 to 1.8V. A V-shaped MnO_2 nanostructure was synthesized[192] by using acetic acid and a microwave-assisted hydrothermal technique. A composite with reduced graphene oxide was prepared using various weight ratios of the oxide and reduced graphene oxide. The specific capacitances of the as-synthesized V-shaped MnO_2 nanostructure and the nanocomposite were 64.75 and 88.95F/g, respectively, at a current density of 0.5A/g in 1M Na_2SO_4 electrolyte. An asymmetrical supercapacitor was constructed by using the nanocomposite as the positive electrode and activated carbon as the negative electrode. The device offered an energy density of 25.14 and 17.95Wh/kg at 0.25 and 1kW/kg power densities, respectively. Nanofibers of α-MnO_2 were combined with N- and S- co-doped reduced graphene oxide by using hydrothermal and ball-milling methods[193]. The oxide nanofibers, with an average diameter of about 40nm, were well-dispersed on the doped graphene oxide nanoflakes. The material was used for supercapacitor electrodes in an electrolyte of N,N-diethyl-N-methyl-N-(2-methoxy-ethyl) ammonium bis (trifluoromethyl-sulfonyl)amide polyvinylidene fluoride-hexafluoropropylene. The device had a specific capacitance of 165F/g at 0.25A/g, with a wide potential window of 0 to 4.5V. The energy-density was 110Wh/kg at a power-density of 550W/kg; with 75% capacitance retention after 10000 cycles at 1A/g.

Mn_2O_3

A nanocomposite of graphene oxide|Mn_2O_3 was prepared[194] on nickel foam by using poly (di-allyl dimethylammonium chloride) as a binder. The resultant electrode had a capacitance of 916.5F/g at a scan-rate 50mV/s, due to its large surface area and porous structure. A specific capacitance of 998.2F/g was measured at a current density of 10A/g during galvanostatic charge/discharge tests, with 91.5% retention after 5000 cycles.

Mn_3O_4

Composites of graphene|Mn_3O_4 were prepared[195] by using coal-sourced graphite rather than natural graphite flakes, and $MnSO_4$ that was transformed into Mn_3O_4 by precipitation in air. Following reduction with hydrazine, graphene-oxide|Mn_3O_4 was used as an electrode for a supercapacitor, with K_2SO_4 as an electrolyte. A maximum specific capacitance of 260F/g was found for a composite with 86%Mn_3O_4 in saturated K_2SO_4 electrolyte solution. The specific energy-density attained 8.7Wh/kg at a current density of 50mA/g when used in a symmetrical supercapacitor; with 92 to 94% retention after 1000 cycles. Reduced-graphene-oxide|Mn_3O_4 composites were prepared[196] by microwave irradiation of the hydrothermal product of reduced-graphene-oxide and $Mn(NO_3)_2$ mixtures. Nanoparticles of Mn_3O_4 with sizes of 20 to 50nm were uniformly distributed on the surface of reduced graphene oxide. The composites had a specific capacitance of 344.8F/g at a current density of 1A/g, using 5M NaOH as an electrolyte. The energy-density of the composites was up to 47.8Wh/kg, with a power-density of 1000W/kg, with 99.2% capacitance retention after 5000 cycles at 1A/g. A composite of Mn_3O_4 nanoparticles and multi-wall carbon nanotubes was used as a positive electrode, reduced graphene oxide was used as a negative electrode, and these were combined with a 1M KOH electrolyte to construct a device having a specific capacitance of 173.36F/g at a 2mV/s scan-rate and an energy-density of 26.8Wh/kg; with 79.3% capacitance retention after 3000 cycles[197].

Graphene oxide with a nanosheet size greater than 20μm, and Mn_3O_4 nanocrystals having a size of about 5nm, were[198] assembled by using a wet-spinning approach and hydrazine-vapor reduction. The use of a suitable proportion of Mn_3O_4 nanocrystals imparted tensile strength and flexibility to the resultant fiber electrode, and also improved the capacitance. Reduced-graphene-oxide|Mn_3O_4-30 fiber electrodes had a maximum volumetric capacitance of 311F/cm^3 at a current density of 300mA/cm^3. The intertwining of 2 as-prepared reduced-graphene-oxide|Mn_3O_4-30 flexible fiber electrodes produced a flexible symmetrical fiber supercapacitor when combined with a polyvinylalcohol-H_3PO_4 gel electrolyte. This device had a maximum volumetric capacitance of 45.5F/cm^3 at a current density of $50mA/cm^3$ and offered a maximum volumetric energy-density of $4.05mWh/cm^3$ with a maximum volumetric power-density of $268mW/cm^3$.

Manganese oxide nanorods with an average diameter 36nm were synthesized[199] by using a hydrothermal method and combined with reduced graphene oxide. The Mn_3O_4 nanorods were dispersed homogeneously within the reduced graphene oxide layers. Nanocomposite electrodes in which the components were in a 1:1 ratio exhibited a good electrochemical performance in 1M Na_2SO_4 or organic electrolytes. Three-electrode measurements indicated a specific capacitance of 228F/g at 5A/g. Two-electrode measurements of symmetrical supercapacitor cells revealed a capacitance of 94F/g, an energy-density of 82Wh/kg at 1A/g, a power-density of 7097W/kg and 95% capacitance retention after 5000 cycles at 5A/g. An asymmetrical supercapacitor based upon Mn_3O_4|reduced-graphene-oxide composite, again with a weight ratio of 1:1, exhibited[200] a relatively good charge-storage ability and a higher specific capacitance and energy-density as compared with those of pure reduced graphene oxide or Mn_3O_4; together with better than 90.3% capacitance retention after 10000 cycles.

Polyaniline was introduced into graphene|manganese-oxide composite in order to increase the specific capacitance[201]. The modified composite was then used as binder-free electrode material in a symmetrical supercapacitor. The specific capacitance of the graphene|Mn_3O_4 composite was increased from 44 to 660F/g by encapsulating polyaniline, with 89% retention after 4000 cycles. The energy-density was 23Wh/kg, with a power-density of 600W/kg, at 1A/g. A symmetrical supercapacitor was based[202] upon Mn_3O_4|reduced-graphene-oxide electrodes and 1M Na_2SO_4 aqueous electrolyte. It maintained a stable working voltage of 2V. Increasing the operational voltage of the device from 1.0 to 2.0V could improve its energy storage ability by more than 20 times. When operating at 2.0V, the device offered an energy-density of 42.2Wh/kg at a power-density of 500.0W/kg.

Graphene|Mn_3O_4 composite was synthesized[203] by means of a flash-irradiation at room temperature which triggered the de-oxygenation of graphene oxide and the decomposition of manganese nitrate. The graphene|Mn_3O_4 positive electrode had an operating potential of 0 to $1.2V_{SCE}$ in aqueous electrolyte. An extended device potential of 2.2V was associated with an energy-density of 18.5Wh/kg at a power-density of 0.28kW/kg. Flower-like Mn_3O_4 and carbon nanohorns were incorporated[204] into 3-dimensional graphene aerogels and used as the positive and negative electrodes, respectively, of an asymmetrical supercapacitor. The device offered an energy-density of $17.4\mu Wh/cm^2$ and a power-density of $14.1mW/cm^2$ ($156.7mW/cm^3$) at 1.4V; with 87.8% capacitance retention after 5000 cycles. A 1-step method was used to prepare[205] reduced-graphene-oxide|Mn_3O_4 composite as the electrodes of a symmetrical supercapacitor, with

$KMnO_4$ being used as the precursor of cubic Mn_3O_4 particles and Na_2SO_3 as the reductant and with graphene oxide being simultaneously reduced. The device had a specific capacitance of 243F/g at 0.5A/g; with 82.3% capacitance retention after 1000 cycles at a current density of 5A/g. It offered an energy-density of 27.92Wh/kg at a power-density of 277.78W/kg.

MnV_2O_6

Graphene|MnV_2O_6 nanocomposites had an optimum ratio of 1:8 and an interconnecting network structure[206]. The maximum specific capacitance was 348F/g at 0.5A/g, with 88% capacitance retention after 3000 cycles at 1A/g. A resultant symmetrical supercapacitor offered a maximum specific energy of 48.33Wh/kg at a specific power of 880.6W/kg.

MoO_2

Graphene|dioxide nanocomposites were prepared[207] hydrothermally which had a specific capacitance of 290F/g and an energy-density of 109.9Wh/kg in an aqueous electrolyte. The composite electrode had a power-density of 5774W/kg at high current-rates.

MoO_3

Nanoplates of MoO_3 were grown onto a graphene and carbon-nanotube framework as a hybrid film[208]. The material had a specific capacitance of 1503F/g at 1A/g and 798.93F/g at a current density of 10A/g; with 96.5% retention after 10000 cycles. An asymmetrical supercapacitor which was constructed by using the MoO_3 composite and graphene-carbon-nanotube as the positive and negative electrode, respectively, had a specific capacitance of 211.71F/g at a current density of 1A/g. The device offered a maximum energy-density of 75.27Wh/kg at a power-density of 816.67W/kg; with 94.2% capacitance retention after 10000 cycles.

MoO_4

A 3-dimensional sulfur-doped-graphene|MoO_4-MoO_2 framework was prepared[209] by using a hydrothermal method and high-temperature calcination. The specific capacity was 1022.7C/g at 1A/g, with 95.2% capacitance retention after 5000 cycles. Porous carbon which was prepared from lotus leaf had a uniform pore structure and flower surface structure which aided charge-storage. An asymmetrical supercapacitor was constructed by using the composite as the positive electrode and lotus-leaf carbon as the negative electrode. This had a specific capacitance of 262.3F/g at a current density of 1A/g, and offered a maximum energy-density of 58.6Wh/kg at a power-density of 801.0W/kg; with 95.0% capacitance retention after 5000 cycles.

Mo_3O_{10}

Ternary composites of reduced-graphene-oxide|Mo_3O_{10}|polyaniline were synthesized[210] by using $Mo_3O_{10}(C_6H_8N)_2\bullet 2H_2O$ and graphene oxide as precursors. When the mass ratio of $Mo_3O_{10}(C_6H_8N)_2\bullet 2H_2O$ to graphene oxide was 8:1, the composite had a maximum specific capacitance of 553F/g in 1M H_2SO_4 and 363F/g in 1M Na_2SO_4 at a scan-rate of 1mV/s. The energy-density attained 76.8Wh/kg at a power-density of 276.3W/kg, and 28.6Wh/kg at a power-density of 10294.3W/kg in H_2SO_4. In the case of Na_2SO_4, the energy-density attained 72.6Wh/kg at a power-density of 217.7W/kg and 13.3Wh/kg at a power-density of 3993.8W/kg; with 86.6 and 73.4% capacitance retention, respectively, after 200 cycles at 20mV/s in 1M H_2SO_4 and Na_2SO_4.

Nb_2O_5

Anodes of Nb_2O_5|reduced-graphene-oxide had a reversible capacitance of 181mAh/g at a current density of 20mA/g, with 85% capacitance retention after 1000 cycles[211]. Mesoporous-carbon|reduced-graphene-oxide cathodes had a capacitance of 110 and 86F/g at a current-density of 500 and 10000mA/g in an organic electrolyte for voltages of 3 to $4.5V_{Li/Li+}$. Supercapacitors with a Nb_2O_5|reduced-graphene-oxide anode and mesoporous-carbon|reduced-graphene-oxide cathode offered a power-density of 25600W/kg at 21Wh/kg; with 82% capacitance retention after 4000 cycles. Oxygen-defect modulated $Ti_2Nb_{10}O_{29-x}$ on interlinked graphene permitted[212] a potential window of up to 1.8V in 1M H_2SO_4 electrolyte and had a capacitance of up to 368.9F/g at 0.5A/g. The oxygen defects in $Ti_2Nb_{10}O_{29-x}$ were capable of stimulating pseudocapacitive behavior and simultaneously suppressing oxygen evolution. A 1.4V $Ti_2Nb_{10}O_{29-x}$-based symmetrical supercapacitor offered a maximum energy-density of $0.58mWh/cm^3$ at a power-density of $0.57W/cm^3$. Orthorhombic Nb_2O_5 was interconnected[213] with reduced graphene oxide nanosheets. Upon combining the composite anode with a nitrogen-doped reduced graphene oxide cathode, the device offered a maximum energy-density of 89Wh/kg at 125W/kg, and an energy-density of 20Wh/kg at 3500W/kg. In order to exploit the advantageous intercalation pseudocapacitive behavior of the oxide and the high conductivity of reduced graphene oxide, a Nb_2O_5|reduced-graphene-oxide composite with an overlapping structure was developed[214]. The composite had a capacitance of 1492F/g at 1A/g, with 88.6% capacitance retention after 3000 cycles in aqueous KOH solution. An asymmetrical supercapacitor with Nb_2O_5|reduced-graphene-oxide and activated-carbon electrodes in aqueous KOH solution and a polyvinylalcohol-KOH gel offered an energy-density of 72Wh/kg at 479W/kg.

(Ni,Co)O

Ostwald ripening was used to synthesize[215] 3-dimensional graphene|metal-organic composite which could be transformed into a graphene|$Ni_xCo_{1-x}O$ aerogel with oxide nanoparticles dispersed on the graphene. The composite had a specific capacitance of 697.8F/g, with 89% capacitance retention at a current density of 20A/g, and 81% retention after 10000 cycles. A supercapacitor which had the composite as the cathode and graphene as the anode offered an energy-density of 27.2Wh/kg at a power-density of 725W/kg; with 86% capacitance retention after 10000 cycles at a current density of 5A/g.

$NiCoO_2$

Porous flower-like-$NiCoO_2$|graphene composite nano-arrays were grown[216] hydrothermally and directly onto nickel foam. The material had a specific capacitance of 1286C/g at 0.5A/g. An asymmetrical supercapacitor which was constructed using $NiCoO_2$/graphene and graphene|Ni-foam electrodes offered an energy-density of 38.5Wh/kg at a 288W/kg power-density; with 86.9% retention after 2000 cycles.

$NiCo_2O_4$

Mesoporous $NiCo_2O_4$ nanoneedles were grown[217] directly onto 3-dimensional graphene-nickel foam by chemical vapor deposition. The material was then used as electrodes for supercapacitor construction. The nanoneedles had a specific capacitance of 1588F/g at 1A/g, with a power-density and energy-density of 33.88Wh/kg at 5kW/kg.

Co-precipitation was used[218] to synthesize a $NiCo_2O_4$|graphene-oxide composite which had a capacitance of 1211.25F/g and 687F/g at a current density of 1A/g and 10A/g, respectively. A resultant asymmetrical supercapacitor with an operational voltage of 1.6V had a specific capacitance of 144.45F/g at a current density of 1A/g. It offered an energy-density of 51.36Wh/kg at 1A/g and a power-density of 50kW/kg at 20A/g, with 88.6% capacitance retention at a current density of 8A/g after 2000 cycles.

Free-standing binder-free electrodes of corrugated $NiCo_2O_4$ nanosheets on N-doped-graphene|carbon-nanotube film had a volumetric capacitance of 482.7F/cm^3 and a gravimetric capacitance of 2292.7F/g at 5A/g; with essentially 100% capacitance retention after 10000 circles at 30A/g. When $NiCo_2O_4$|graphene|carbon-nanotube composite was used[219] as the cathode and graphene|carbon-nanotube composite was used as the anode, the device offered energy-densities of 42.71Wh/kg at 775W/kg and 24.69Wh/kg at a power-density of 15485W/kg. It exhibited a volumetric energy-density of 25.90Wh/l and a volumetric power-density of 9389W/l.

Graphene was decorated with mesoporous rod-like $NiCo_2O_4$ by using a hydrothermal method[220], leading to a capacitance of 845F/g. An asymmetrical supercapacitor was constructed by using this nanocomposite as the positive electrode and doped carbon as the negative electrode. The device offered an energy-density of 52.2Wh/kg, with 97.3% capacitance retention after 10000 cycles.

A layered $NiCo_2O_4$|reduced-graphene-oxide nanocomposite was prepared[221] via the layer-by-layer assembly of exfoliated Co-Ni and graphene-oxide nanosheets having opposite charges; followed by freeze-drying and annealing. This arrangement prevented re-stacking of the reduced graphene oxide and the aggregation of $NiCo_2O_4$ nanosheets. The nanocomposite had a specific capacity of 1388F/g at 0.5A/g, with a rate performance of 840F/g at 30A/g; with 90.2% retention after 20000 cycles at 5A/g. A resultant asymmetrical supercapacitor with $NiCo_2O_4$|reduced-graphene-oxide composite and activated-carbon as the positive and negative electrodes, respectively, offered an energy-density of 57Wh/kg at a power-density of 375W/kg; with a working potential of 0 to 1.5V.

A mesoporous $NiCo_2O_4$-nanorod|graphene-oxide and N-doped graphene were prepared[222] hydrothermally and used as the electrodes of a supercapacitor. The $NiCo_2O_4$|graphene-oxide composite electrode had a specific capacitance of 709.7F/g at a current density of 1A/g, with 84.7% capacitance retention at 6A/g after 3000 cycles. A resultant high-voltage asymmetrical supercapacitor had $NiCo_2O_4$|graphene-oxide composite and N-doped graphene as the positive and negative electrodes, respectively, in 1M KOH aqueous electrolyte. The device offered an energy-density of 34.4Wh/kg at a power-density of 800W/kg, and 28Wh/kg at a power-density of 8000W/kg; with 94.3% capacitance retention at 5A/g after 5000 cycles.

A 3-dimensionally cross-linked Co_3O_4|$NiCo_2O_4$|N-doped-graphene composite, grown onto nickel foam, has been prepared[223] by hydrothermal reaction and calcination. When used as an electrode for a supercapacitors, the material had a specific capacitance of 2430F/g at 1A/g, with 93.7% retention after 5000 cycles. A resultant asymmetrical supercapacitor had a specific capacitance of 118.6F/g at a current density of 1A/g, with a maximum energy-density of 52.9Wh/kg and power-density of 544W/kg. The energy-density was up to 23.9Wh/kg at a maximum power-density of 3630W/kg.

A 3-dimensional graphene hydrogel, embedded with nickel foam, has been used[224] as a high surface-area support. Following the electrodeposition of $NiCo_2O_4$ nanoflakes, the support was used as a supercapacitor electrode. A $NiCo_2O_4$|graphene-hydrogel|Ni-foam composite electrode had an hierarchical open-pore structure. An as-prepared electrode had a capacitance of 3.84F/cm^2 at 2mA/cm^2, with 71.6% retention at 50mA/cm^2. A resultant asymmetrical supercapacitor which had $NiCo_2O_4$|graphene-hydrogel|Ni-foam as the positive electrode and graphene-hydrogel|Ni-foam as the negative electrode offered a maximum energy-density and power-density of 65Wh/kg and 18.9kW/kg, respectively, with 92% capacitance retention after 5000 cycles.

Nanowires of $NiCo_2O_4$ have been grown[225] onto 3-dimensional porous graphene aerogel by 2-step hydrothermal means. The material had a specific capacitance of 720F/g at a current density of 1A/g, with 84% capacitance retention after 1000 cycles. A resultant asymmetrical supercapacitor had $NiCo_2O_4$|graphene-aerogel as the positive electrode and activated-carbon as the negative electrode. The device offered an energy-density of 25.41Wh/kg at a power-density of 658W/kg; with 78% capacitance retention after 3000 cycles.

The usual nanosheet-like structure of $NiCo_2O_4$ tends to aggregate, leading to blocked electrochemically-active sites. After distributing[226] the oxide nanosheets over N-doped graphene frameworks, the specific capacitance was 1198F/g at 1A/g. A device that involved $NiCo_2O_4$|N-doped-graphene|activated-carbon offered a much higher energy-density than did one which involved $NiCo_2O_4$|activated-carbon.

A 3-dimensional elastic sponge which comprised single-walled carbon nanotubes and graphene was prepared[227] by flame-burning within 20s. Hierarchical $NiCo_2O_4$ nanosheets were deposited onto the above scaffold by using hydrothermal methods and annealing. Due to the 3-dimensional structure the compressed electrode material had a capacitance of 2050F/g or 1200F/cm^3; with essentially 100% capacitance retention after 6800 cycles at a current density of 20A/g, and 73.2% capacitance retention when the current density was increased from 2 to 80A/g. A resultant asymmetrical supercapacitor, with $NiCo_2O_4$|carbon-nanotubes|graphene-sponge as the positive electrode and activated polyaniline-derived carbon as the negative electrode, offered energy-densities of 53.5Wh/kg or 26.9Wh/l; with 83.5% capacitance retention after 10000 cycles. A 3-dimensional nitrogen-doped single-wall carbon nanotube and graphene elastic sponge composite with a density of 0.8mg/cm^3 was prepared[228] by the pyrolysis of carbon nanotubes and graphene-oxide coated polyurethane sponge. Compressed $NiCo_2O_4$|Ni(OH)$_2$| graphene| carbon-nanotube electrode material had a capacitance of 1810F/g or 847.7F/cm^3 at 1A/g. A resultant asymmetrical supercapacitor, with the above material as the cathode and graphene|carbon-nanotubes as the anode, offered an energy-density of 54Wh/kg at 799.9W/kg and 37Wh/l at 561.5W/l.

Template-free low-temperature solvothermal synthesis of porous-graphene|$NiCo_2O_4$ nanorod composites was followed[229] by calcination in air; thus producing a high-porosity nanocomposite which permitted fast ion diffusion, together with good mechanical strength. The capacitance was 1533F/g at a scan-rate of 5mV/s, and 1684F/g at a current density of 1A/g; with 94% capacitance retention after 10000 cycles at a current density of

8A/g in 2M KOH electrolyte. A resultant supercapacitor offered an high energy-density of 45.3Wh/kg and a power-density of 17843.5W/kg.

Positive reduced-graphene-oxide|NiCo$_2$O$_4$|MnO$_2$ and negative reduced-graphene-oxide|Bi$_2$O$_3$|Cu$_2$S electrodes have been combined[230] with polyvinylalcohol-KOH with added K$_3$[Fe(CN)$_6$] in order to eliminate the drawbacks of aqueous-based asymmetrical supercapacitors. The specific capacitance was 2107F/g and 1617F/g at 5mV/s for the positive and negative electrodes, respectively. The resultant asymmetrical devices permitted potentials of up to 2.1V and offered an energy-density of 60.52Wh/kg at a power-density of 625W/kg; with 95.17% capacity retention after 10000 charge/discharge cycles.

Large-area NiCo$_2$O$_4$|reduced-graphene-oxide composites with a hierarchical structure were prepared[231] by 1-step ultrasonic spraying onto nickel foam and used as binder-free electrodes for supercapacitors with aqueous KOH electrolyte. The electrodes had a specific capacitance of 871F/g at a current density of 1A/g, with complete capacitance retention after 30000 cycles. The resultant asymmetrical supercapacitor offered a maximum energy-density of 29.3Wh/kg at a power-density of 790.8W/kg; again with complete capacitance retention after 30000 cycles at 20A/g.

A 3-dimensional self-supporting flexible structure has been created[232] with NiCo$_2$O$_4$ nanowires anchored on a graphene framework. Electrodes made from the graphene|NiCo$_2$O$_4$ had a specific capacitance of 1118F/g at 1A/g; with 95.04% retention after 5000 cycles. A resultant symmetrical supercapacitor had a specific capacitance as 270F/g at 1A/g, and offered an energy-density of 9.37Wh/kg at 250W/kg. A NiCo$_2$O$_4$|/graphene-quantum-dot composite[233] had a specific capacitance of 481.4F/g at 0.35A/g.

Nanosheets of NiCo$_2$O$_4$ were prepared[234] by solvothermal reaction and heat-treatment. In a 3-electrode system, it had a maximum specific capacitance of 978F/g at 1.5A/g. A free-standing partially-reduced-graphene-oxide|carbon-nanotube film was obtained, by vacuum filtering and solvothermal means, which had a maximum specific capacitance of 236F/g at 0.5A/g. A resultant asymmetrical supercapacitor with partially-reduced-graphene-oxide|carbon-nanotube|NiCo$_2$O$_4$ had a specific capacitance of 82F/g at 0.5A/g, an energy-density of 18.8Wh/kg at 0.32kW/kg and 99.5% capacitance retention after 5000 cycles at 5A/g.

NiFe$_2$O$_4$

A composite[235] which comprised NiFe$_2$O$_4$ nanocubes anchored on reduced graphene oxide cryogel had a capacitance of 488F/g at a constant current density of 1A/g, with

89.8% capacitance retention after 10000 cycles. A resultant device offered an energy-density of 62.5Wh/kg, with 93.2% capacitance retention after 6000 cycles.

NiMn₂O₄

A ternary composite was based upon $NiMn_2O_4$, reduced graphene oxide and polyaniline[236]. The specific capacitance was 757F/g at a current density of 1A/g. The composite had a maximum energy-density of 70Wh/kg, with about 93% capacitance retention after 2000 cycles. Three-dimensional self-assembled $NiMn_2O_4$|CoS core-shell microspheres were synthesized[237] by using hydrothermal and electrodeposition means. The microspheres were composed of many nanoflakes having a diameter of about 1.8μm. The composites had a specific capacitance of up to 1751F/g at a current density of 1A/g, and 1270F/g at a current density of 30A/g; with 95% capacitance retention after 5000 cycles at 10A/g. An asymmetrical supercapacitor with composite and graphene electrodes offered an energy-density of 44.56Wh/kg at a power-density of 700.51W/kg and a power-density of 20.99kW/kg at 29.1Wh/kg; with 94% capacitance retention after 5000 cycles at 10A/g.

NiMoO₄

A hydrothermal method was used to synthesize[238] hexahedral α-MnMoO₄ and a composite with graphene. Pseudocapacitive MnMoO₄ and the MnMoO₄|graphene composite in 1M Na_2SO_4 had maximum specific capacitances of 234F/g and 364F/g, respectively, at a current density of 2A/g. Maximum energy-densities of 130Wh/kg and 202.2Wh/kg were found, respectively, for the MnMoO₄ and the MnMoO₄|graphene composite at a power delivery-rate of 2000W/kg; with 88% capacitance retention after 1000 cycles at 8A/g.

Nickel molybdate nanowires were coated[239] onto chemical-vapor-deposited 3-dimensional graphene skeletons. The binder-free graphene|α-NiMoO₄ composite was used as the positive electrode in a supercapacitor and had a specific capacitance of 1194F/g at 12mA/cm²; with 87.3% capacitance retention after 2000 cycles. The energy-density was 41Wh/kg at a power-density of 1319W/kg.

Reduced-graphene-oxide|NiMoO₄ nanocomposite was synthesized[240] by using a hydrothermal method, followed by annealing at 450C. The rod-like NiMoO₄, with a length of 1 to 5μm, was spread over the reduced graphene-oxide surface. In a 3-electrode configuration, the composite had a capacitance of 680F/g, with 68% capacitance retention after 4000 cycles at 3A/g. In a 2-electrode configuration, the electrode had a capacitance of 74F/g and offered an energy-density of 26.3Wh/kg at 0.75A/g; with 56.7% capacitance retention after 4000 cycles at 1.5A/g. A 9V supercapacitor had a capacitance and energy-density of 1.09F/g and 12.3Wh/kg, respectively at 8mA; with 72% capacitance retention after 1500 cycles.

Core-shell $NiMoO_4|V_2CT_x|$reduced-graphene-oxide composite was produced[241] by using a room-temperature ionic-liquid assisted hydrothermal method. Bamboo-shaped $MoO_2|Fe_2O_3|$N-doped-carbon was used as a negative electrode in an asymmetrical supercapacitor with $NiMoO_4|V_2CT_x|$reduced-graphene-oxide. The supercapacitor offered an energy-density of 56.1Wh/kg at 800W/kg; with about 90.7% capacitance retention after 5000 cycles.

A composite was developed[242] which combined the advantages of the high specific capacitance of $NiMoO_4$ with the high rate-capability of Co_3O_4. The composite exhibited a pseudocapacitive performance of about 1722.3F/g at a current density of 1A/g, with 91% capacitance retention after 6000 cycles. An asymmetrical supercapacitor was constructed by using the composite and activated-carbon as the positive and negative electrode, respectively. This offered an energy-density of 37.1Wh/kg at a power-density of 798.0W/kg; with 100% capacitance retention after 4000 cycles.

A wrinkled-nanosheet structure with a mixture of $NiMoO_4$ and triethylamine was created hydrothermally[243]. The combined ions of Ni^{2+}/Ni^{3+} and Mo^{2+}/Mo^{3+} markedly increased energy storage. The $NiMoO_4$ was deposited onto nickel foam with no binder, and the electrode had specific capacitance of 415.66F/g at 8mA/cm^2.

Manganese-doped $NiMoO_4$ has smaller unit-cell parameters and is more reactive than $NiMoO_4$ because of the defects created by the doping. A composite which consisted of $Mn_{0.1}Ni_{0.9}MoO_4$ mesoporous nanorods and reduced graphene oxide was combined[244] with alkaline polyvinylalcohol as an electrolyte. The composite had a specific capacitance of 109.3F/g at 1A/g, with 96.1% capacitance retention after 200 cycles, and offered an energy-density of 49.2Wh/kg at 1800W/kg. The supercapacitor was highly flexible and retained 83.6% of its specific capacitance when bent. When tested in a 3-electrode system, the $Mn_{0.1}Ni_{0.9}MoO_4|$reduced-graphene-oxide composite had a maximum specific capacitance of 688.9F/g at 0.5A/g.

NiO

A 1-pot chemical reduction method was used[245] to prepare NiO and reduced graphene oxide composites. Composite NiO|reduced-graphene-oxide electrodes with a weight-ratio of 6:4 had a maximum specific capacitance of 461F/g and an energy-density of 36.0Wh/kg at 0.21A/g in 6M KOH electrolyte. A 1-step electrochemical method was used[246] to form NiO-quantum-dot|graphene-flake composites. The 3nm NiO quantum dots were uniformly deposited onto few-layer graphene surfaces. The composite had a specific capacitance of 1181.1F/g at a current density of 2.1A/g. When used in an

Materials Research Forum LLC
https://doi.org/10.21741/9781644901939

asymmetrical supercapacitor, the device offered an energy-density of 27.3Wh/kg at 1562.6W/kg.

A ternary system which comprised graphene, Ni and NiO was used[247] as the positive electrode in a supercapacitor and had a specific capacitance of 1410F/g at 1A/g and 1020F/g at 15A/g. Porous carbon was used as a negative electrode. When these electrodes were combined, the asymmetrical supercapacitor had a maximum specific capacitance of 183.8F/g and an energy-density of 65.3Wh/kg. At a power-density of 8000W/kg, the device could maintain a level of 42.2Wh/kg. It retained a specific capacitance of 120F/g after 3000 cycles at 8A/g. An asymmetrical supercapacitor was constructed[248] from Ce-doped NiO as the positive electrode and reduced graphene oxide as the negative electrode in aqueous KOH electrolyte. Oxide which contained 1%Ce had a high specific capacitance. The supercapacitor had a specific capacitance of 110F/g at a scan-rate of 5mV/s and offered a maximum energy-density of 26.27Wh/kg, with 91.6% retention after 1000 cycles.

A layered NiO|reduced-graphene-oxide composite was prepared[249] in which 2-dimensional hydroxide and graphene-oxide nanosheets exhibited nanoscale dispersion with high interfacial interaction. The as-prepared material had a specific capacitance of 782F/g at 0.5A/g, with 94.1% retention after 3000 cycles at 2A/g. An asymmetrical supercapacitor which was constructed with NiO|reduced-graphene-oxide as the cathode and activated-carbon as the anode offered a maximum energy-density of 32.5Wh/kg at a power-density of 375W/kg and 19.78Wh/kg at 7500W/kg, with 92.7% retention after 3000 cycles. An asymmetrical supercapacitor was based[250] upon graphene|NiO core-shell nanosheets as the positive electrode and 3-dimensional drilled graphene sheet hydrogel as the negative electrode. The graphene|NiO electrode had a specific capacitance of 1092F/g at 1A/g. Because of its mesoporous hydrogel structure, the porous drilled graphene electrode had a specific capacitance of up to 268F/g at 1A/g. The optimized device had a specific capacitance of 122.5F/g at 0.5A/g and offered a maximum energy-density of 43.2Wh/kg, with 95.5% capacitance retention after 5000 cycles at 6A/g.

Free-standing graphene-paper|NiO and graphene-paper|Ni composites were prepared[251] in which NiO nanoclusters and Ni nanoparticles were encapsulated by graphene sheets. The nanoclusters and nanoparticles impeded re-stacking of the graphene sheets and provided space for rapid ion diffusion and electron transport. At a current density of 0.5A/g, the graphene-paper|NiO and graphene-paper|Ni electrodes had specific capacitances of 306.9 and 246.1F/g, with 98.7% and 95.6% retention, respectively, after 10000 cycles. The device itself exhibited 93.7% retention after 10000 cycles. Nanoparticles of NiO and NiFe$_2$O$_4$ were grown[252] on 3-dimensional N-doped graphene by calcination and microwave synthesis. The composite had a specific capacitance of 1556.5F/g at a current-density of 1A/g in 6M KOH solution, 92.5% capacitance retention after 2000 cycles. The ternary composite offered a maximum energy-density of 55.6Wh/kg, with a power-density of 1598.7W/kg.

Nanoparticles of NiO were uniformly embedded in a spherical carbon matrix, grown onto reduced graphene oxide via the pyrolysis[253] of a nickel metal organic framework precursor on graphene oxide. The composite electrode had a specific capacity of 496C/g at a current density of 1A/g. A supercapacitor which was constructed using the composite electrode, and a porous carbon electrode, offered an energy-density of 35.9Wh/kg at a power-density of 749.1W/kg, with essentially 100% retention after 3000 cycles. Supercapacitor electrodes were made[254] from a sulfonated graphene substrate, well-dispersed NiO and polyaniline. The composite had a specific capacitance of 1350F/g at a current density of 1A/g, with 92.23% retention after 5000 cycles. An asymmetrical supercapacitor which was constructed using composite and activated-carbon electrodes had a specific capacitance of 308.8F/g and offered an energy-density of 109.8Wh/kg and a power-density of 0.8kW/kg; with 91.15% retention after 10000 cycles. Graphene-oxide nanosheets and 2-dimensional $Ni(OH)_2$ nanosheets were interleaved[255] and heat-treated to yield 2-dimensional NiO|reduced-graphene-oxide composites in which NiO nanosheets were homogeneously dispersed on the surface of reduced-graphene-oxide nanosheets. The material had a capacitance of 343C/g at 1A/g. Electrodes made from the composite were used to construct a symmetrical supercapacitor which offered an energy-density of up to 5.4Wh/kg at a power-density of 0.43kW/kg; with 90% capacitance retention after 10000 cycles at 10A/g.

$NiSnO_3$

A composite, $NiSnO_3$|graphene, was prepared[256] hydrothermally with the formation of elongated particles having a size of about 6nm. The surface area of the composite was $162m^2/g$, while that of plain $NiSnO_3$ nanoparticles was $101m^2/g$. The maximum specific capacitance of the composite was 891F/g at a scan-rate of 5mV/s, while that of $NiSnO_3$ was 570F/g at the same scan-rate. The improvement in the performance of the composite was attributed to graphene incorporation, which provided a high surface area. An asymmetrical supercapacitor which was constructed by using the nanocomposite and activated-carbon as positive and negative electrodes, respectively, had an operating potential of 0 to 0.8V. The device offered an energy-density of 42.54Wh/kg at a power-density of 0.34kW/kg, with 88.3% capacitance retention after 4000 cycles.

$Ni_3V_2O_8$

Graphene|$Ni_3V_2O_8$ nanocomposites were synthesized[257] solvothermally, and a weight-ratio of 1:4 was found to be optimum. It had a specific capacitance of 748F/g at a current density of 0.5A/g, and offered an energy-density of 103.94Wh/kg at a power-density of 45.61kW/kg; with 71% capacitance retention after 3000 cycles at 0.5A/g.

RuO_2

An asymmetrical supercapacitor has been made[258] by using reduced graphene oxide sheets which were modified with RuO_2 or polyaniline as the anode and cathode, respectively. The device exhibited a much better capacitance than that of a symmetrical supercapacitor made using either of the modified graphenes alone as the electrodes. This improvement was attributed to a wider potential window in an aqueous electrolyte, leading to an energy-density of 26.3Wh/kg. A power-density of 49.8kW/kg was obtained at an energy-density of 6.8Wh/kg. Flexible asymmetrical supercapacitors have been based[259] upon using ionic-liquid modified graphene film as the negative electrode and hydrous RuO_2-containing composite film as the positive electrode, with a polyvinylalcohol-H_2SO_4 electrolyte. The devices were optimized so as to have a maximum cell voltage of up to 1.8V, and offer an energy-density of 19.7Wh/kg at a power-density of 6.8kW/g. They could operate at a rate of 10A/g, with 79.4% capacitance retention after 2000 cycles. Three-dimensional hydrous RuO_2-anchored graphene and carbon nanotube hybrid foam have been used as supercapacitor electrodes[260]. That is, the graphene foam was covered with networks of RuO_2 nanoparticles of less than 5nm, and anchored carbon nanotubes. Devices which were based upon these materials had a specific capacitance of 502.78F/g and an areal capacitance of 1.11F/cm^2, leading to an energy-density of 39.28Wh/kg and a power-density of 128.01kW/kg.

Ruthenium oxide plus reduced graphene oxide nanoribbon composite, with 72.5wt%RuO_2, had a specific capacitance of up to 677F/g at a current density of 1A/g when measured in a 3-electrode array with 1M H_2SO_4 electrolyte[261]. There was 91.8% capacitance retention rate at 20A/g. A symmetrical supercapacitor which was based upon electrodes made of the composite offered an energy-density of 16.2Wh/kg at a power-density of 9885W/kg. Ruthenium and carboxylated graphene were co-electrodeposited and the composite was electro-oxidized in aqueous 0.5M H_2SO_4 to yield electrodes for supercapacitor use[262]. The electrodes exhibited a specific capacitance of 756F/g and an energy-density of 101Wh/kg at a power-density of 2.5kW/kg.

Ruthenium oxide was coated onto graphene-coated Cu foil by electroplating and used to construct flexible supercapacitor electrodes[263]. Their properties were measured by using a 3-electrode array in 0.5M H_2SO_4 electrolyte. The graphene insertion-layer was a key factor in improving the structural and electrochemical properties of the electrode film under bending conditions. The electrode had a specific capacitance of 1561F/g

Materials Research Forum LLC

https://doi.org/10.21741/9781644901939

(0.015F/cm) at a scan-rate of 5mV/s, with 98% retention when bent. The flexible RuO_2/graphene/Cu electrodes offered an energy-density of about 13Wh/kg at a power-density of about 21kW/kg. Ruthenium oxide and graphene nanocomposites for supercapacitor electrodes were prepared[264] by using hydrothermal methods which evenly dispersed the oxide over the graphene sheets. Contact between the nanoparticles and the graphene improves the conductivity of the RuO_2 and prevented re-stacking and agglomeration of the graphene. The RuO_2|graphene composite had a specific capacitance of 441.1F/g at 0.1A/g, with 94% capacitance retention after 1000 cycles. The composites gave a voltage window of 1.6V, with an energy-density of 61.2Wh/kg at a power-density of 183.8W/kg for 1M Na_2SO_4 electrolyte. Ruthenium oxide and graphene composites were directly grown onto Ni foam by cyclic voltammetric deposition[265]. A sandwich structure was formed from a mixed solution of graphene oxide and ruthenium trichloride. Symmetrical aqueous supercapacitors which were based upon RuO_2|graphene electrodes permitted an operational voltage window of 1.5V. An energy-density of 43.8Wh/kg at a power-density of 0.75kW/kg was observed, with 39.1Wh/kg being retained at a power-density of 37.5kW/kg; with 92.8% capacitance retention after 10000 cycles.

SiO_2

Ternary composites of reduced-graphene-oxide|silica|polyaniline were prepared[266] and used to construct a symmetrical supercapacitor (coin) cell with the composite as the electrodes. The energy-density of the composite was 10Wh/kg at a power-density of 2310W/kg; with 75.2% capacitance retention after 6000 cycles at 0.8A/g. The reduced-graphene-oxide|silica|aniline composite was further improved[267] by sulfonating the polyaniline. The specific capacitance was increased from 24 to 780F/g by the sulfonation.

Sm_2O_3

A graphene aerogel was cross-linked, by p-phenylenediamine composite, to Sm_2O_3 nanoparticles[268]. The p-phenylenediamine provided a high surface area by reducing the adhesion of graphene ultra-thin sheets. The cross-linked structure of the nanocomposite enhanced its supercapacitive behavior in 6M KOH. The specific capacitance of the nanocomposite electrode attained 591F/g at 5mV/s; with 92.7% retention after 4000 cycles. The composite electrode increased the energy-density to up to 55Wh/Kg.

SnO_2

Ternary electrode material, based upon graphene, SnO_2 and polypyrrole, was obtained[269] by 1-pot synthesis. The nanocomposite consisted of a thin conducting film of polypyrrole on the surface of graphene|SnO_2. A specific capacitance of 616F/g was measured 1mV/s in 1M H_2SO_4. The electrode exhibited no obvious decay after 1000 cycles at 1A/g. The specific power-density and energy-density could attain 9973.26W/kg and 19.4Wh/kg, respectively. A similar ternary material was based[270] upon graphene, SnO_2 and poly(3,4-ethylene-dioxythiophene. A maximum specific capacitance of 184F/g in 1M H_2SO_4 and 180F/g in 1M Na_2SO_4, respectively, was observed. The energy-density attained 22Wh/kg at a power-density of 238.3W/kg, and 17.1Wh/kg at a power-density of 5803.3W/kg in H_2SO_4. In the case of Na_2SO_4, the energy-density attained 23.4Wh/kg at a power-density of 253W/kg, and 10.2Wh/kg at a power-density of 3684W/kg, respectively; with 100% and 70% capacitance retention at 1A/g after 5000 cycles in 1M H_2SO_4 and Na_2SO_4, respectively. Oxide|graphene nanocomposites were prepared[271] by using a wet chemical method. Tetragonal SnO_2 with a particle size of about 50nm was uniformly distributed on graphene sheets. The electrochemical performance was studied in 6M KOH electrolyte. A maximum specific capacitance of 818.6F/g was found for SnO_2|graphene composite at a 5mV/s scan-rate. Supercapacitors which operated at high temperatures were constructed[272] by combining SnO_2-doped graphene aerogel electrodes and ionic-liquid composite electrolytes. Tiny SnO_2 nanoparticles were chemically bonded to the graphene sheets. A combination of porous aerogel electrode, and composite electrolyte led to a specific capacitance of 541F/g and an energy-density of 160Wh/kg at 125C. Tin oxide quantum dots, graphene oxide and polypyrrole were deposited[273] on titanium foil as a positive electrode while graphene oxide charcoal on titanium foil was used as a negative electrode, separated by a polyvinylalcohol-KOH gel-electrolyte. The device had a maximum specific capacitance of 1296F/g, and offered an energy-density of 29.6Wh/kg and a maximum power-density of 5310.26W/kg; with 90% capacitance retention after 11000 cycles.

$SrFe_{12}O_{19}$

Uniform strontium ferrite nanorods were grown[274] on graphene by using a surfactant assisted hydrothermal method. The $SrFe_{12}O_{19}$-nanorods|graphene composite had a specific capacitance of 681.2 and 450.9F/g at 1 and 20A/g, respectively; with 95.5% capacitance retention after 10000 cycles at 1A/g. They offered an energy-density and power-density of 47.3Wh/kg at 500.9W/kg and 31.3Wh/kg at 10247.7W/kg.

TiO_2

Titania nanobelts and nanoparticles were combined[275] with reduced graphene oxide to form nanocomposites for supercapacitor electrodes. The specific capacitance of reduced-graphene-oxide|TiO_2 composites was higher than that of monolithic reduced-graphene-oxide, TiO_2 nanoparticles or nanobelts and the optimum performance corresponded to a reduced-graphene-oxide:TiO_2 mass ratio of 7:3. The specific capacitances of reduced-graphene-oxide|TiO_2-nanobelts and reduced-graphene-oxide|TiO_2-nanoparticles at the mass ratio of 7:3 were 225F/g and 62.8F/g, respectively, at a current density of 0.125A/g.

A supercapacitor was constructed[276] from an anode of anatase TiO_2|reduced-graphene-oxide composite and a cathode of activated-carbon in a non-aqueous electrolyte. Over a voltage range of 1.0 to 3.0V, the device offered 42Wh/kg of energy at 800W/kg. An energy-density of up to 8.9Wh/kg was possible at a 4s charge/discharge rate. Reduced-graphene-oxide|TiO_2 nanocomposites were synthesized[277] in various weight ratios: 1:1, 1:2, 1:5, 1:10, by using a microwave-assisted method. Symmetrical supercapacitors were constructed from electrodes of these materials. The as-prepared symmetrical device had a specific capacitance of 524.02F/g at 2mV/s for a weight ratio 1:5 and an energy-density of 50.07Wh/kg at 2mV/s for a weight ratio of 1:1 and a power-density of 58.6kW/kg at a scan-rate of 1000mV/s; with 6.6% retention after 1000 cycles. A membrane-like flexible supercapacitor was constructed[278] which had a reduced-graphene-oxide|TiO_2|graphene-oxide|reduced-graphene-oxide|TiO2 sandwich structure. The reduced-graphene-oxide|TiO_2 layer was the active material and the middle graphene-oxide layer was the separator. This supercapacitor offered a volumetric capacity of 237F/cm^3 and a volumetric energy-density of 16mWh/cm^3.

Uniform TiO_2 nanowires on reduced graphene oxide nanosheets were synthesized[279] by using a hydrothermal method. The composite had a capacitance of 202.5F/g at 1A/g, with 81.9% capacitance retention after 5000 cycles. The TiO_2 nanowires prevented agglomeration, shortened the ion diffusion length and facilitated charge transfer at the interfaces and within the electrodes. A symmetrical supercapacitor which was constructed by using the nanocomposite had a specific capacitance of 45.7F/g and offered an energy-density of 9.08Wh/kg at a power-density of 598W/kg. A laser-irradiation method was used[280] to modify the electrical conductivity of TiO_2 and form an oxygen-deficient-TiO_2|graphene composite. Oxygen vacancies increased the intrinsic electrical conductivity, and the graphene sheets facilitated superficial electron-transfer. A symmetrical supercapacitor which was constructed by using electrodes made from the composite offered a maximum energy-density of 14.1Wh/kg and a maximum power-density of 8.5kW/kg.

VO$_2$

An oxide|reduced-graphene-oxide heterostructure was used as an anode, and activated-carbon on carbon cloth was used as a cathode, to construct a device[281]. The mixed valency of ions within the as-prepared oxide matrix allowed redox reactions to occur at a low potential, producing a graphene|oxide anode with a working potential of 0.01 to -3$V_{Li/Li+}$. Sheet-on-sheet heterostructured graphene|oxide had a specific capacitance of 1214mAh/g at 0.1A/g after 120 cycles. The overall device had a maximum gravimetric energy-density of 126.7Wh/kg and a maximum gravimetric power-density of about

10000W/kg within an operating voltage range of 1 to 4V. It permitted rapid charging-times of 5 and 834s, with energy-densities of 15.6 and 82Wh/kg, respectively.

V_2O_5

Flexible paper electrodes were prepared[282], without binders, from a V_2O_5 nanowire and graphene composite. Coin-type supercapacitors were constructed by using the composite paper as the anode and spectracarb-fiber cloth as the cathode. The composite electrode offered an energy-density of 38.8Wh/kg at a power-density of 455W/kg. A maximum power-density of 3.0kW/kg was found for a current discharge rate of 5.5A/g. A layer-by-layer assembly technique was developed[283] in which a graphene layer was alternately inserted between multiwalled carbon-nanotube films which were coated with 3nm of V_2O_5. The insertion of a conductive graphene spacer between the carbon-nanotube films coated with V_2O_5 prevented agglomeration between the nanotube films and increased the specific capacitance to up to 2590F/g. Layer-by-layer supercapacitor electrodes offered an energy-density of 96Wh/kg at a power-density of 800W/kg, with better than 97%, capacitance retention after 5000 cycles. Porous V_2O_5|graphene aerogels were produced[284] by means of a sol-gel method and the *in situ* growth of V_2O_5 nanofibers on graphene sheets. Supercapacitors which were based upon these composites offered a specific capacitance of 486F/g and an energy-density of 68Wh/kg. A composite of V_2O_5 nanofibers and exfoliated graphene was prepared[285] via *in situ* growth of the fibers. The oxide layer, uniformly grown on the graphene surface, had a high specific surface area and easy ion-conduction paths. The composite had a specific capacitance of 218F/g at a current density of 1A/g, an energy-density of 22Wh/kg and a power-density of 3594W/kg. An asymmetrical supercapacitor was constructed by using activated-carbon cloth and the composite as positive and negative electrodes, respectively. The device had a capacitance of 279F/g at 1A/g, an energy-density of 37.2Wh/kg and a power-density of 3743W/kg.

Free-standing graphene|V_2O_5 monolithic composites have been prepared[286] by using a hydrothermal process. Flexible graphene sheets served as a binder to connect belt-like V_2O_5 into a 3-dimensional network. The composite could be shaped into flexible film. A specific capacitance of 358F/g was found for graphene|V_2O_5 monolith, as compared with that (272F/g) for graphene|V_2O_5 flexible film in a 0.5M K_2SO_4 solution. An asymmetrical supercapacitor was constructed by using composite monolith as the positive electrode and graphene monolith as the negative electrode. The device could be reversibly charged/discharged at a cell voltage of 1.7V in 0.5M K_2SO_4. The asymmetrical capacitor offered an energy-density of 26.22Wh/kg at a power-density of 425W/kg, with 90% capacitance retention after 1000 cycles at a current density of 5A/g. A ternary nanocomposite, based upon graphene oxide, polypyrrole and V_2O_5, was prepared[287] by using 1-step electrochemical deposition onto a stainless-steel substrate and tested in 0.5M Na_2SO_4 solution. The nanocomposite had a specific capacitance of 750F/g at a current density of 5A/g; with 83% retention after 3000 cycles. A symmetrical supercapacitor which was constructed using nanocomposite electrodes offered a maximum energy-density of 27.6Wh/kg at a power-density of 3600W/kg, and a maximum power-density of 13680W/kg at an energy-density of 22.8Wh/kg.

Three-dimensional V_2O_5|graphene|multiwalled-carbon-nanotube aerogels were synthesized[288] by using a sol-gel method: the V_2O_5 grew along the surfaces of the carbon nanotubes and the graphene. Two types of 1-dimensional fiber, V_2O_5 nanofibers and V_2O_5-coated carbon nanotubes, and 2-dimensional sheets were assembled into a 3-dimensional porous island-chain structure. The composite had a specific capacitance of 504F/g and an energy-density of 70Wh/kg; with 82.9% capacitance retention after 32500 cycles. Hydrothermally synthesized[289] 3-dimensional orthorhombic V_2O_5 nanosheets and V_2O_5|graphene-foam composites were produced. An optimized V_2O_5|graphene-foam-150mg composite had a specific capacity of 73mAh/g. This was much higher than that (60mAh/g) of pristine V_2O_5 nanosheets at a current of 1A/g. A supercapacitor was constructed by combining a carbon-based negative electrode, with the above optimum composite as a positive electrode, in 6M KOH electrolyte. The device offered an energy-density of 39Wh/kg at a power-density of 947W/kg at a specific current of 1A/g in a voltage window of 0.0 to 1.6V; with 74% capacitance retention after 10000 cycles at a specific current of 10A/g. A V_2O_5-nanowire|reduced-graphene-oxide composite was prepared[290] which had a gravimetric capacitance of 1002F/g, with 83.94% capacitance retention after 5000 cycles. The energy-density was 116Wh/kg at a power-density of 1520W/kg.

Nanocomposites of V_2O_5 and variously-doped graphene sheet were prepared[291] by using a solvothermal method, and their supercapacitance properties were determined in a KOH electrolyte. A maximum specific capacitance of 1032.6F/g was measured in the case of V_2O_5|N-doped-graphene at a 1mV/s scan rate. The energy-density was 185.86Wh/kg at a power-density of 37.20W/kg. A layer-by-layer-assembled reduced-graphene-oxide|V_2O_5 heterostructure was patterned[292] with interdigitated electrodes, deposited directly onto a flexible current-collector, in order to form a supercapacitor. The interdigitated electrodes offered free access to the electrolyte ions. A resultant flexible in-plane microsupercapacitor had a capacitance of 24mF/cm^2 and 34.28F/cm^3, and offered an energy-density of 3.3μWh/cm^2 and 4.7mWh/cm^3; with 93.7% capacitance retention after 10000 cycles. A reduced-graphene-oxide|V_2O_5|polyaniline nanocomposite, with weight percentages of 5.88, 11.76 and 82.36%, respectively, was used to construct[293] a supercapacitor which offered an energy-density of 54.62 Wh/kg and a power-density of 1636.5 W/kg at a current density of 1A/g. The maximum specific capacitance was 273F/g.

V_3O_7

A highly porous $V_3O_7 \bullet H_2O$-nanobelt|carbon-nanotube|reduced-graphene-oxide ternary composite[294] having a 3-dimensional microstructure had a specific surface area of up to 53.7m^2/g. The ternary composite had a specific capacitance of 685F/g at 0.5A/g and an energy-density of 34.3 Wh/kg, with 99.7% capacitance retention after 10000 cycles.

WO_3

A solvothermal method has been used to synthesize[295] graphene-supported oxide nanowires for use as an electrode. The graphene|WO_3 nanowire nanocomposite had a specific capacitance of 465F/g at 1A/g, with 97.7% capacitance retention after 2000 cycles in 0.1M H_2SO_4 electrolyte. An asymmetrical supercapacitor was constructed by using the above nanocomposite as the negative electrode and activated-carbon as the positive electrode. The device offered an energy-density of 26.7 Wh/kg at a 6kW/kg power-density, and could retain 25 Wh/kg at a 6kW/kg power-density after 4000 cycles. A 3-dimensional hollow graphene foam was prepared[296] via template-assisted chemical vapor deposition, with monoclinic WO_3 interconnected nanoparticles and 2-dimensional $Ti_3C_2T_x$ sheets bilaterally loaded onto the inner and outer sides of hollow graphene foam. The 3-dimensional free-standing WO_3|$Ti_3C_2T_x$|graphene electrode had a specific capacitance of 573F/g at 5mV/s, with 93.3% retention after 5000 cycles. An asymmetrical supercapacitor had a specific capacitance of 145.2F/g (111.3mF/cm^2) at 5mV/s, an energy-density of 27.2 Wh/kg (20.83μWh/cm^2) at a power-density of 752 W/kg and 93% retention after 10000 cycles. An asymmetrical supercapacitor was constructed[297] by using monoclinic WO_3 nanoplates as a negative electrode and highly reduced graphene oxide as a positive electrode, with 1M H_2SO_4. The device had a maximum specific capacitance of 389F/g at a current density of 0.5A/g, with an associated energy-density of 93 Wh/kg at a power-density of 500 W/kg, with 92% capacitance retention after 5000 cycles.

WV_2O_7

Cumulative assembly of WV_2O_7 nanorods led to 2-dimensional nanosheets of interwoven tungsten vanadate on graphene[298]. When the as-prepared material was used in a supercapacitor with H_2SO_4 electrolyte, the average specific capacitance was 346.4F/g. The maximum energy-density was up to 27.8 Wh/kg at 950 W/kg, while the power-density was 23.8kW/kg at 16.2 Wh/kg.

$ZnCo_2O_4$

Ternary $ZnCo_2O_4$|reduced graphene oxide|NiO-nanowire arrays were grown[299] directly onto nickel foam by using hydrothermally assisted thermal annealing. The Ni-foam was served successively as NiO precursor, binder and current collector. The resultant 3-dimensional composite had a specific capacitance of 1256F/g at a current density of 3A/g in 6M KOH solution, with about 80% capacitance retention after 3000 cycles. The maximum energy-density was 62.8Wh/kg, at a maximum power-density of 7492.5W/kg.

A binder-free supercapacitor electrode was based[300] upon $ZnCo_2O_4$ nanoflakes and N-doped reduced graphene oxide. The morphology of the oxide was tailored by changing the substrate from bare nickel foam to reduced graphene oxide-supported Ni-foam and then to N-doped reduced graphene oxide-supported Ni-foam. The latter electrode had a specific capacitance of 1613F/g at 1A/g, with 97.3% retention after 5000 cycles. An asymmetrical supercapacitor which was based on this composite had an energy-density of 36.5Wh/kg.

$ZnMn_2O_4$

Graphene nanoribbons embedded with $ZnMn_2O_4$ 7nm nanospheres were used[301] as electrodes in a device, and a gel polymer membrane was used as the electrolyte. The uniform dispersion of $ZnMn_2O_4$ led to enhanced transport of the electrolyte ions. A resultant $ZnMn_2O_4$|graphene-nanoribbon electrode supercapacitor had a maximum operating potential of 2.7V and offered an energy-density of about 37Wh/kg and a power-density of about 30kW/kg at 1.25A/g.

$(Zn,Ni)MoO_4$

A graphene|$Zn_{1-x}Ni_xMoO_4$ composite was synthesized[302], in which x was 0, 0.2, 0.4, 0.6 or 0.8, and tested in 2M KOH electrolyte. The $Zn_{0.6}Ni_{0.4}MoO_4$ exhibited superior behavior. The material exhibited little aggregation, and interconnected layers of graphene with interspersed rice-like grains of the molybdate. It offered an energy-density and power-density of 62.3Wh/kg and 448.5W/kg, respectively, with a specific capacitance of 555.5F/g at 1A/g and 85% retention after 5000 cycles.

ZnO

Ternary composites of the form, reduced-graphene-oxide|ZnO|poly(p-phenylenediamine), have been synthesized[303] by using a hydrothermal-polymerization method. The reduced-graphene-oxide sheet substrate and thin poly(p-phenylenediamine) coating prevented the volume expansion of ZnO during the charging/discharging. When the mass ratio of the components was 1:8:8, the composite had a specific capacitance of 320F/g at 5mV/s. A supercapacitor which used this material as the electrodes offered a maximum energy-density of 18.14Wh/kg and a maximum power-density of 10kW/kg. A composite was made[304] which consisted of ZnO and NiO on a nitrogen-doped 3-dimensional graphene network. The oxide nanoparticles had a size of 10 to 20nm. When used as an electrode material in a supercapacitor, the nanocomposites had a specific capacitance of 1839.4F/g at a current density of lA/g in 6M KOH solution; with some 93% retention after 6000 cycles at 10A/g. The ZnO|NiO|graphene offered a maximum energy-density of 35.32Wh/kg, with a power-density of 139.72W/kg at a current density of 10A/g in a 2-electrode system. A binder-free nanocomposite which consisted of ZnO nanoparticles, grown directly onto graphene sheets by electrospraying, was used[305] as an electrode in supercapacitors. The ZnO particles grew in a uniformly distributed manner on the graphene sheets, with negligible agglomeration. A symmetrical supercapacitor which was constructed by using this composite had an energy-density of 67mWh·/cm^3 and a power-density of 6000mW/cm^3; with 90% capacitance retention after 1000 cycles.

ZnV_2O_6

Composites of the form, ZnV_2O_6|polypyrrole, were synthesized[306] by using hydrothermal methods and low-temperature *in situ* oxidative polymerization. The material had a specific capacitance of 723.6F/g at 1A/g. An asymmetrical supercapacitor which had the composite as the positive electrode and nitrogen-doped reduced graphene oxide as the negative electrode offered a maximum specific capacitance of 109.2F/g at 1A/g. The device offered a maximum energy-density of 34Wh/kg at a power-density of 748.7W/kg, with 93% capacitance retention after 3000 cycles.

$Zn_3V_2O_8$

Graphene|$Zn_3V_2O_8$ nanocomposites were prepared[307] by using solvothermal methods and were tested in 2M KOH aqueous electrolyte. The composite had a sheet-on-sheet nanostructure which intertwined to form a 3-dimensional network. Nanocomposites having a 1:3 weight-ratio exhibited a specific capacitance of 564F/g at 0.8A/g. A resultant symmetrical supercapacitor offered an energy-density of 78Wh/kg at a power-density of 75.5kW/kg.

SULFIDES

Al_2S_3

Graphene-oxide supported sulfide composite was prepared hydrothermally[308] and had a specific capacitance of 1687.97F/g at a scan-rate of 5mV/s and a specific capacitance of 2178.16F/g at a current-density of 3mA/cm^2. The energy-density was 108.91Wh/kg at a current-density of 3mA/cm^2, and a power-density of 978.92W/kg was found at a current-density of 15mA/cm^2; with 57.84% capacitance retention after 1000 cycles.

CoMoS$_2$

Vertically-aligned Co_3S_4|$CoMo_2S_4$ ultra-thin nanosheets on reduced graphene oxide were prepared hydrothermally[309]. The Co_3S_4 was well-matched to the $CoMo_2S_4$ surface. Sulfide sheets with a thickness of 3 to 5nm were uniformly dispersed on the reduced graphene-oxide, thus increasing the contact area with the electrolyte. The material had a capacitance of 1457.8F/g at 1A/g, with 97% capacitance retention after 2000 cycles. An asymmetrical supercapacitor was constructed by using the composite as the positive electrode and activated-carbon as the negative electrode. This device offered an energy-density of 33.1Wh/kg at a power-density of 0.85kW/kg; with 93.8% retention after 5000 cycles. Sulfide nanocrystals were deposited on 3-dimensional sulfur-doped graphene and tested in KOH and Na_2SO_4 electrolytes[310]. The composite had a specific capacitance of 1288.33F/g in the KOH electrolyte and 991.67F/g in the Na_2SO_4 electrolyte at a current density of 1A/g; with 90% capacitance retention after 2000 cycles. An asymmetrical supercapacitor was constructed with the composite as the positive electrode and S-doped graphene as the negative electrode. With a KOH electrolyte, the device offered a maximum energy-density of 38.73Wh/kg at a power-density of 0.81kW/kg. In a Na_2SO_4 electrolyte, the energy-density was up to 47.32Wh/kg at a power-density of 0.94kW/kg.

CoNi$_2$S$_4$

Nanocomposites of $CoNi_2S_4$|graphene have been synthesized[311] by using solvothermal methods and used as supercapacitor electrodes. The latter had a specific capacitance of 1621F/g at 0.5A/g, with 76.7% capacitance retention at 10A/g, and no capacitance loss after 2500 cycles at 5A/g. When used as the positive electrode of an asymmetrical supercapacitor, the material had a specific capacitance of 126.6F/g at 0.5A/g, with 87.4% capacitance retention after 5000 cycles. The highest energy-density was 39.56Wh/kg at a power-density of 374.8W/kg. Reduced graphene oxide-coated $CoNi_2S_4$ was grown onto nickel foam substrates by using a 2-step method[312]. The reduced-graphene-oxide coating tended to improve the cyclability of the sulfide electrode without sacrificing capacitance. The $CoNi_2S_4$|reduced-graphene-oxide electrode had a specific capacitance of 1680F/g at 1A/g, with 62% retention at 1 to 20A/g. When used as the positive electrode in an asymmetrical supercapacitor, together with a porous-carbon negative electrode, the device offered an energy-density of 51.7Wh/kg at 762W/kg, with 60.5% capacitance retention after 5000 cycles. Reduced-graphene-oxide|$CoNi_2S_4$|NiCo-double-hydroxide composite with a p-n junction structure was created[313] by depositing n-type nickel-cobalt layered double hydroxide around p-type reduced-graphene-oxide|$CoNi_2S_4$. The resultant electrode had a specific capacitance of 1310C/g at 1A/g, with 77% retention after 5000 cycles. A supercapacitor which was based upon the p-n junction battery electrode offered an energy-density of 57.4Wh/kg at 323W/kg.

CoS

A 3-dimensional porous structure was based upon carbon-dot supported sulfide-decorated graphene oxide hydrogel[314]. The carbon dots acted as a stabilizer for the CuS nanoparticles and strongly bound the sulfide nanoparticles to graphene oxide within the 3-dimensional hydrogel structure. The composite had a specific capacitance of 920F/g at a current density of 1A/g. The composite was used as the positive electrode of an asymmetrical supercapacitor, with reduced graphene oxide as the negative electrode. The device offered a maximum energy-density of up to 28Wh/kg, with 90% capacitance retention after 5000 cycles. A reduced-graphene-oxide|CuS composite was prepared[315] solvothermally in which the sulfide microstructure consisted of regular tiny nanoparticles supported by graphene sheets. The composite had a maximum specific capacitance of 946F/g at 10mV/s, and 906F/g at 1A/g; with 89% retention after 5000 cycles at 5A/g. The energy-density was 105.6Wh/kg at a power-density of 2.5kW/kg. A 3-dimensional porous-graphene|CoS_2|Ni_3S_4 composite electrode was synthesized[316] by using solvothermal and vulcanization methods. It had specific capacitance of 1739F/g at a current density of 0.5A/g. Core-shell cobalt sulphide|manganese molybdate composites have been prepared[317] on reduced graphene oxide|Ni-foam by using hydrothermal and electrodeposition methods. When the CoS mass was 0.4mg/cm^2, the composite had a specific capacitance of 3074.5F/g at 1A/g, with 87% retention after 5000 cycles. An asymmetrical device which was based upon the composite offered a maximum energy-density of 50.3Wh/kg at 415.8W/kg; with 96% retention after 8000 cycles. A hydrothermal method was used[318] to prepare nitrogen-doped carbon dots, flower-like sulfide and reduced graphene oxide nanosheets. A supercapacitor which was constructed by using flower-like CoS|N-doped-carbon-dot composite as the cathode and reduced-graphene-oxide|N-doped-carbon-dots as the anode offered an energy-density of about 36.6Wh/kg at 800W/kg; with 85.9% retention after 10000 cycles at 10A/g. Porous cobalt sulfide (CoS_x)|reduced-graphene-oxide films were used[319] as a flexible free-standing supercapacitor electrode, following the assembly and sulfidation of 2-dimensional metal organic framework nanoflakes and graphene oxide. Zeolitic imidazolate-67 nanocubes were added to an aqueous solution of graphene oxide to produce a mixed dispersion of imidazolate-67 and graphene oxide, and also induce the reduction of graphene oxide. An asymmetrical supercapacitor was constructed by using reduced-graphene-oxide|CoS_x-graphene|reduced-graphene-oxide film and activated-carbon as the positive and negative electrode, respectively. The device offered an energy-density of 10.56Wh/kg and a power-density of 2250W/kg; with 92.8% capacitance retention after 10000 cycles.

Co_3S_4

A 3-dimensional Ni-foam|graphene|Co_3S_4 composite film was used[320] as an electrode for a supercapacitor. The sulfide film consisted of small-diameter hollow nanospheres that were clustered together. The electrodes exhibited 97.8% capacitance retention after 8000 cycles.

Nickel-foam|reduced-graphene-oxide-hydrogel|Ni_3S_2 and Ni-foam|reduced-graphene-oxide-hydrogel|Co_3S_4 composites were synthesized[321] by using a 2-step hydrothermal method. Secondary hydrothermal treatment led to the growth of porous Ni_3S_2 nanorods and a Co_3S_4 nanosheet on a Ni|reduced-graphene-oxide substrate. When Ni|reduced-graphene-oxide|Ni_3S_2 and Ni|reduced-graphene-oxide|Co_3S_4 were combined with 6M KOH electrolyte, they had specific capacitances of 987.8 and 1369F/g, respectively, at 1.5A/g; 97.9% and 96.6% capacitance retention, respectively, at 12A/g after 3000 cycles. An aqueous asymmetrical supercapacitor was constructed by using as-prepared Ni|reduced-graphene-oxide|Co_3S_4 as the positive electrode and Ni|reduced-graphene-oxide|Ni_3S_2 as the negative electrode. The device offered energy-densities of 55.16 and 24.84Wh/kg at power-densities of 975 and 13000W/kg, respectively; with 96.2% capacitance retention after 3000 cycles at 12A/g. Binder-free graphene-nanosheet-wrapped Co_3S_4 electrodes were prepared[322] on Ni-foam by using a 2-step hydrothermal method. The Co_3S_4|reduced-graphene-oxide electrode had a specific surface area of 30m2/g, and the electrode had an areal and specific capacitance of 8.33F/cm^2 and 2314F/g, respectively. The electrode charged within 30s, while the energy-density remained at 54.32Wh/kg at a power-density of 6.25kW/kg; with 92.6% retention after 1000 cycles. An asymmetrical supercapacitor was constructed by using Co_3S_4 and Co_3S_4|reduced-graphene-oxide as the positive and negative electrode, respectively. The device had an areal capacitance of about 164mF/cm^2 with 89.56% retention after 5000 cycles. It offered an energy-density of 1.09Wh/kg at a power-density of 398W/kg, and a 0.31Wh/kg energy-density could be retained at a power-density of 750W/kg.

Co_9S_8

The Co_9S_8 nanoparticles were deposited onto 3-dimensional graphene by using a glucose-assisted hydrothermal method[323]. Treatment in 1M KOH solution increased the surface roughness of the composite electrode. The latter had a specific capacitance of 1721F/g and exhibited great cycling stability at a current density of 16A/g. An asymmetrical supercapacitor which had the composite, and reduced graphene oxide hydrogel, as the electrodes offered an energy-density of 31.6Wh/kg at a power-density of 910W/kg; with

86% capacitance retention after 6000 cycles. Island-like sulfide nanoparticles, comprising 2-dimensional Co_9S_8 nanosheets, were dispersed evenly on the surface of 2-dimensional N-doped graphene[324]. The nanocomposite had a specific capacitance of 2992F/g at 1A/g. An asymmetrical supercapacitor which was constructed by using the composite as the positive electrode and activated-carbon as the negative electrode offered an energy-density of 88.81Wh/kg at a power-density of 0.63kW/kg, a power-density of 21.80kW/kg at an energy-density of 30.30Wh/kg, and 80.61% capacitance retention after 10000 cycles at a current-density of 10A/g.

Cu_2NiSnS_4

Quaternary chalcogenide Cu_2NiSnS_4 nanoparticles were grown[325] *in situ* on 2-dimensional reduced graphene oxide. An as-fabricated device offered an areal capacitance of 655.1mF/cm², while a volumetric capacitance of 16.38F/cm³ was found at a current density of 5mA/cm², combined with an energy-density of 5.68mW/cm³; comparable to that of lithium thin-film batteries. The device also retained 89.2% of the initial capacitance after 2000 cycles.

MnS

Porous manganese sulfide nanocrystals, anchored on graphene oxide, were obtained[326] by using a hydrothermal method which was based upon the Kirkendall effect. The honeycomb-like sulfides (40-80nm) and the 3-dimensional sandwich structure led to a specific capacitance of 390.8F/g at 0.25A/g, with 78.7% retention at 10A/g after 2000 cycles. An asymmetrical supercapacitor which had the composite as the positive material and activated-carbon as the negative electrode, had a specific capacitance of 73.63F/g, an energy-density of 14.9Wh/kg at 66.5W/kg and 12.8Wh/kg at a power-density of 4683.5W/kg. Mesoporous γ-MnS nanosheets with surface sulfur vacancies were prepared[327] on 3-dimensional reduced graphene oxide substrates. The composites had a specific capacitance of 27.98F/cm³ at 0.03A/cm³. The surface S vacancies increased the number of charge-carrier adsorption sites and speeded up the charge-carrier transfer process. A flexible asymmetrical supercapacitor which had the composite as the positive electrode and N-doped reduced graphene oxide as the negative electrode offered an energy-density of 2.783Wh/cm³ at a power-density of 0.057W/cm³; with 88.5% capacitance retention after 5000 cycles at 5A/g.

MoS_2

Layered molybdenum disulfide and graphene composites have been synthesized[328] by using an L-cysteine-assisted solution method. The layered MoS_2-graphene coalesced into a 3-dimensional sphere-like architecture. The maximum specific capacitance of the MoS_2|graphene electrodes attained 243F/g at a discharge current density of 1A/g. The energy-density was 73.5Wh/kg at a power-density of 19.8kW/kg, with 92.3% capacitance retention after 1000 cycles at a current density of 1A/g. Polyaniline|MoS_2 nanocomposites were prepared via *in situ* polymerization and used as supercapacitor electrodes[329]. The layered nanostructure provided a larger contact surface area for the intercalation and de-intercalation of protons in or out of active materials, and shortened the path-length for electrolyte-ion transport. The maximum specific capacitance was 575F/g at 1A/g. An energy-density of 265Wh/kg was obtained at a power-density of 18.0kW/kg; with 98% capacitance retention after 500 cycles at a current density of 1A/g, Layered MoS_2 was deposited, by using microwave heating, onto reduced graphene oxide at 3 concentrations[330]. The first layers of the sulfide were directly bonded to the oxygen of oxide by covalent chemical bonds. The electroactive material could be reversibly cycled between 0.25 and 0.8V in 1M $HClO_4$ solution at low concentrations of MoS_2, and between 0.25 and 0.65V at medium or high concentrations MoS_2 layers on graphene. The specific capacitance at 10mV/s was 128, 265 and 148F/g for low, medium and high concentrations of MoS_2, respectively. The energy-density was 63Wh/kg, with 92% capacitance retention after 1000 cycles. Molybdenum disulfide and nanosheet-graphene composites were prepared[331], with *in situ* reduction of graphene oxide. The composite had a specific capacitance of 270F/g at a current density of 0.1A/g in a neutral aqueous electrolyte. The energy-density of the composite electrode was 12.5Wh/kg at a power-density 2500W/kg. A dip-and-dry method was used to stack few-layer MoS_2 nanosheets onto a 3-dimensional graphene network[332]. The assembled MoS_2 nanosheets had an expanded interlayer spacing of about 0.75nm and were stacked discontinuously on the surface of the graphene. The composite capacitance exhibited 76.73% capacitance retention upon increasing the current density from 1 to 100A/g. An asymmetrical supercapacitor was constructed by using the composite and activated-carbon as electrodes. The device had a working voltage window of 1.6V, together with power- and energy-densities of 400.0 to 8001.6W/kg and 36.43 to 1.12Wh/kg, respectively.

Nanocomposites of polyaniline and an equal wt% of graphene and MoS_2 were prepared[333] via the *in situ* oxidative polymerization of polyaniline or polyaniline-graphene. These were well-dispersed materials, with an interleaved structure of graphene and MoS_2 that encapsulated polyaniline nanorods. Ternary composite electrodes had a specific capacitance of 142.30F/g, with 98.11% capacitance retention. Polyaniline|graphene|MoS_2 symmetrical electrodes offered an energy-density of 2.65Wh/kg at a power-density of 119.21W/kg. An electrode which was based upon hollow MoS_2 sphere intercalated graphene film was used[334] as an electrode in a supercapacitor, where the structure provided conducting channels that promoted electrolyte penetration and exploited the surface area as much as possible. The resultant binder-free electrode had a specific capacitance of 286.8F/g. A supercapacitor which was based upon MoS_2|grapheme and activated-carbon offered a maximum energy-density of 22.0Wh/kg at 800W/kg. Ternary heterostructures which consisted of reduced graphene oxide, molybdenum disulfide and tungsten disulfide were prepared[335], such that a MoS_2|WS_2 heterostructure was uniformly created on the conductive graphene support. The reduced-graphene-oxide|MoS_2|WS_2 had a surface area of $109m^2$/g within an hierarchical porous architecture. When used as a supercapacitor electrode, the reduced-graphene-oxide|MoS_2|WS_2 exhibited pseudocapacitive behavior in KOH solution. The specific capacitance was 365F/g at 1A/g. When used as the positive electrode in an asymmetrical supercapacitor, the maximum energy-density was 15Wh/kg at a power-density of 373W/kg; with about 70% capacitance retention after 3000 cycles. Layered MoS_2 could be combined with 3-dimensional graphene so as to form a vertical cross-linked structure[336]. The MoS_2 nanosheets were vertically loaded onto the inner and outer surfaces of graphene when the concentration of MoS_2 was 0.20mg/l. The specific capacitance of such a composite attained $2182.33mF/cm^2$ at a current density of $1mA/cm^2$, and remained after 5000 cycles. When the current density was increased from 1 to $100mA/cm^2$, the specific capacitance remained at 78.9%. Hybrid energy storage devices could offer an energy-density of $130.34Wh/m^2$.

Conductive MoS_2|reduced-graphene-oxide nanosheets were immobilized on carbon cloth by using a hydrothermal method[337]. The intimate contact between the nanostructure and the cloth maintained the stability of the structure. A supercapacitor electrode which was made from the material had a specific capacitance of 331F/g at 0.75A/g, with complete capacitance retention after 15000 cycles. An asymmetrical supercapacitor which was made by using it offered an energy-density of 29.2Wh/kg and a power-density of 4517.7W/kg. A flexible asymmetrical supercapacitor was constructed[338] by using a polyaniline|poly(3,4-ethylene-dioxythiophene|polyaniline|reduced-graphene-oxide tetralayer film as the positive electrode and a poly(3,4-ethylene-dioxythiophene|MoS_2 film as the negative electrode on a flexible polyethylene terephthalate substrate with a polyvinylalcohol-H_2SO_4 gel electrolyte. The tetralayers were coated onto the polyethylene terephthalate by using a layer-by-layer self-assembly method, whiles the other electrode was coated onto polyethylene terephthalate by using a drop-coating technique. The asymmetrical supercapacitor offered a maximum energy-density of $5.4mWh/cm^3$ at a power-density of $110mW/cm^3$. It retained an energy-density of $4.0mWh/cm^3$ at a power-density of $265mW/cm^3$ for a cell voltage of 0.8V.

A sulfide|graphene-nanoribbon framework for flexible supercapacitors was based[339] upon laser-induced MoS_2|MnS|graphene frameworks which had an areal specific capacitance of $58.3mF/cm^2$ at $50\mu A/cm^2$, an areal energy-density of $7.0\mu Wh/cm^2$ at $50\mu A/cm^2$ and an areal power-density of $49.9\mu W/cm^2$ at $50\mu A/cm^2$; with 93.6% capacitance retention after 10000 cycles. Supercapacitor electrodes were prepared from nanocomposites of MoS_2|reduced-graphene-oxide|polypyrolle-nanotubes[340]. The supercapacitor electrodes were then modified by 100MeV heavy-ion bombardment at fluences of up to $10^{13}/cm^2$. As a result of the bombardment, the specific capacitance, energy-density and cycling stability were improved up to a fluence of $3.3 \times 10^{12}/cm^2$, but then degraded at the highest fluence. The specific capacitance of 1561F/g of a pristine nanocomposite electrode was increased to 1875F/g by bombardment. The bombarded electrodes were more stable, with 91% retention after 10000 cycles, than were pristine electrodes, with 72% retention.

Binder-free free-standing MoS_2-nanosphere|reduced-graphene-oxide paper was assembled into a flexible symmetrical capacitor[341]. The capacitance of the device was 323F/g at a current density of 0.2A/g, with a maximum energy-density of 44.9Wh/kg; with 76.8% retention after 500 cycles. Upon bending device, a capacitance of 277F/g could be maintained, with 73.7% retention after 500 cycles, and an energy-density of 38.5Wh/kg. Hydrothermally-assembled fibers which comprised a graphite core and a MoS_2 nanosheet graphene-oxide sheath were used[342] as electrodes for supercapacitors. The optimum composite fiber, with 34.9wt%MoS_2, had a volumetric capacitance of $421F/cm^3$ at a can-rate of 5mV/s and a 51.0% capacitance retention when the scan-rate was increased from 2 to 100mV/s. The core-sheath fiber permitted rapid reversible redox kinetic, and its surface capacitive energy storage contributed 75 to 80% of the total energy storage. A supercapacitor based on the fiber electrodes offered a volumetric capacitance of $94F/cm^3$ at $0.1A/cm^3$ and an energy-density of $8.2mWh/cm^3$ at a power-density of $40mW/cm^3$.

A 1-pot chemical method was used to prepare MoS_2 and MoS_2|graphene nanostructures which had specific capacitances of 175 and 756F/g, respectively, at 0.5A/g; and retained 88% of the original capacitance after 10000 cycles[343]. The maximum energy-density of the MoS_2|graphene supercapacitor was 26.6Wh/kg at a power-density of 125W/kg. Laser direct writing was used to prepare MoS_2 nanoparticle-embedded laser-induced graphene patterned electrodes on various substrates[344]. The embedding of MoS_2 nanoparticles tended to improve the hydrophobic properties of MoS_2|laser-induced-graphene electrodes. A flexible planar supercapacitor had an areal capacitance of $35.3mF/cm^2$ and an energy-density of $4.91mWh/cm^2$ at a power-density of $0.18mW/cm^2$. The addition of quantum-dots can markedly improve the binding state of MoS_2 and N- and S-doped graphene and increase the overall defect density[345]. When the quantum-dot dosage was 20mg, the quantum-dot-MoS_2|co-doped-graphene composite had a specific capacitance of 564.3F/g at a scan-rate of 10mV/s. At a current density of 20A/g, the capacitance retention rate was 92.8% after 10000 cycles. An asymmetrical supercapacitor was constructed by using the co-doped-graphene and the quantum-dot-MoS_2|co-doped-graphene composite. When the power-density was 900W/kg, the energy-density attained 68.8Wh/kg. At a current density of 20A/g, the specific capacitance retention-rate was 92.2% after 10000 cycles.

Nano-urchin MoS_2|VS_2 was grown homogeneously onto reduced graphene oxide by using a hydrothermal method and tested in an aqueous electrolyte system[346]. Graphene|MoS_2|VS_2 material had a specific capacitance of 460F/g ($287.5F/cm^2$) at 0.5A/g and 310F/g ($193.75F/cm^2$) at 10A/g. A symmetrical configuration offered an energy-density of 49Wh/kg ($67.86mWh/cm^2$) at a power-density of 700.25W/kg. Three-dimensional graphene oxide decorated monodispersed hollow urchin γ-MnS had a capacitance of 858F/g at 1A/g. A flexible asymmetrical supercapacitor device was constructed[347] by having surface-activated carbon cloth decorated with γ-MnS|graphene as the positive electrode and 3-dimensional graphene on carbon cloth as the negative electrode. This device offered an energy-density of 26Wh/kg at a power-density of 500W/kg at 1A/g, and an energy-density of 17.8Wh/kg at a power-density of 1500W/kg at 3A/g. Vertically-aligned MoS_2-nanosheet|polypyrrole composites on reduced graphene oxide were prepared hydrothermally[348]. Ultra-thin MoS_2 nanosheets, mixed with polypyrrole lamellae, were coated onto the surface of reduced graphene oxide so as to form a ternary nanostructure. The conductivity of MoS_2 was improved by the polypyrrole and graphene. The composite electrode had a capacitance of 1942F/g (215.8mAh/g) at a density of 1A/g. An asymmetrical supercapacitor, constructed using these materials, offered an energy-density of 39.1Wh/kg at a power-density of 0.70kW/kg.

Uniform small MoS_2 nanoparticles could be grown rapidly onto graphene, thus imparting as-prepared composites with better capacitive properties[349]. The resultant MoS_2|graphene composites had a specific capacitance of 401.1F/g at 1A/g, with 95.0% capacitance retention after 10000 cycles at 5A/g. They also offered an energy-density of 26.4Wh/kg at a power-density of 1015.4W/kg. A composite of MoS_2 and N,P-doped graphene nanoflowers was prepared hydrothermally[350], and had a specific capacitance of 588F/g at 1A/g. A symmetrical supercapacitor which used the composite for both electrodes offered an energy-density of 24.34Wh/kg and 91.67% capacitance retention at 3A/g after 5000 cycles. A reduced-graphene-oxide|MoS_2|poly(3,4-ethylenedioxythiophene) composite was deposited onto carbon-fiber cloth by hydrothermal polymerization[351]. The graphene oxide acted as an oxidant to initiate the polymerization of 3,4-ethylenedioxythiophene, and the graphene oxide then transformed into reduced graphene oxide. The flexible electrode had an areal capacitance of $241.81mF/cm^2$ at a current density of $0.5mA/cm^2$; with 93.7% capacitance retention after 5000 cycles. Molybdenum disulfide and poly(3,4-ethylenedioxythiophene) imparted a high pseudocapacitance. A supercapacitor which was based upon the ternary electrodes offered an energy-density of $1.44\mu Wh/cm^2$ at a power-density of $0.058mW/cm^2$. Ultra-thin MoS_2-covered MoO_2 nanocrystal arrays on a S-doped 3-dimensional graphene framework were obtained by hydrothermal processing and high-temperature annealing[352]. The specific capacitance was up to $1150.37F/g$, with 94.6% capacitance retention after 10000 cycles. The addition of FeS_2 nanoflowers further increased device performance. An asymmetrical supercapacitor with a positive electrode made from MoS_2-MoO_2|graphene and a negative electrode made from FeS_2|graphene, worked efficiently at a voltage of 1.7V and offered an energy-density of $87.38Wh/kg$ at a power-density of $683.94W/kg$.

NiCo₂S₄

A hydrothermally produced[353] $NiCo_2S_4$|reduced-graphene-oxide composite had a specific capacitance of $2003F/g$ at $1A/g$ and $1726F/g$ at $20A/g$, with 86.0% capacitance retention from $1A/g$ to $20A/g$ and 86.0% retention after 3500 cycles. An asymmetrical supercapacitor was constructed by using the above composite as the positive electrode and activated-carbon as the negative electrode in 2M KOH electrolyte. This supercapacitor offered an energy-density of $21.9Wh/kg$ at a power-density of $417.1W/kg$, and an energy-density of $13.5Wh/kg$ at a power-density of $2700W/kg$. A 3-dimensional core/shell structure[354] comprised $NiCo_2S_4$ nanotubes as the core and $Co_xNi_{(3-x)}S_2$ nanosheets as the shell. The as-synthesized graphene|$NiCo_2S_4$|$Co_xNi_{(3-x)}S_2$ composite electrode had an areal capacitance of $15.6F/cm^2$ at a current density of $10mA/cm^2$; with 93% capacitance retention after 5000 cycles. An asymmetrical device, made by using graphene|$NiCo_2S_4$|$Co_xNi_{(3-x)}S_2$ as the positive electrode and reduced graphene oxide as the negative electrode, offered an energy-density of $23.9Wh/kg$ and a power-density of $2460.6W/kg$ at an operating current of 100mA. A solvothermal method was used to prepare[355] $NiCo_2S_4$ nanoparticles and nanosheets, immobilized on a graphene aerogel, with the microstructure of the sulfide being markedly affected by the pH of the solution. Uniformly dispersed sulfide nanoparticles with a diameter of 25.94nm were present on the graphene when the solution pH was 8.2. The $NiCo_2S_4$|graphene-aerogel electrode had a specific capacitance of $704.34F/g$ at $1A/g$; with 80.3% capacitance retention 1500 cycles at $2A/g$. An asymmetrical supercapacitor which was based upon the above electrode offered an energy-density of $20.9Wh/kg$ at a power-density of $800.2W/kg$.

One method for the construction of $NiCo_2S_4$ heterostructures was based[356] upon electrochemically exfoliated graphene, the growth of a bimetallic precursor and its *in situ* conversion into $NiCo_2S_4$ nanoparticles. This led to a capacitance of $1802.5F/g$ at $1A/g$ and 86.1% retention at $10A/g$. Nitrogen-doped mesoporous carbon spheres were used as the anode in devices. A resultant supercapacitor offered an energy-density of $30.4Wh/kg$ at $800.0W/kg$, with 80.1% capacitance retention after 9000 cycles. The incorporation of electroactive Fe_2O_3 and $NiCo_2S_4$ into graphene aerogel yielded[357] composites that could be used as negative and positive electrodes, respectively, in a KOH aqueous electrolyte. A specific capacitance of 200F/g (178C/g) and 386F/g (170C/g) was found for graphene-aerogel|Fe_2O_3 (20wt% oxide) and graphene-aerogel| $NiCo_2S_4$ (77wt% sulfide), respectively. A supercapacitor which was constructed with graphene-aerogel|Fe_2O_3 as the negative electrode and graphene-aerogel| $NiCo_2S_4$ as the positive electrode offered a specific capacitance of 93F/g (128C/g) at 0.1A/g, corresponding to a maximum energy-density of 25Wh/kg at a power-density of 54W/kg; with 72.3% capacitance retention after cycling at 2A/g for 5000 cycles. High-porosity reduced-graphene-oxide|$NiCo_2S_4$ nanocomposite, prepared[358] by using a 1-step microwave-assisted method, consisted of $NiCo_2S_4$ nanoneedles with diameters of some nanometres, anchored to the surface of the graphene sheets. As-prepared electrodes had a specific capacitance of 1320F/g at 1.5A/g, with better than 96% retention after 2000 cycles. A resultant asymmetrical supercapacitor which used the reduced-graphene-oxide|$NiCo_2S_4$ as the cathode and reduced graphene oxide as the anode had a capacitance of 146F/g at 1A/g and offered an energy-density of 46.7Wh/kg at a power-density of 1200.8W/kg, with an operating voltage of 0.8V. A nanocomposite[359] consisting of $NiCo_2S_4$ nanoparticles in a reduced graphene oxide matrix had a specific capacitance of 963 to 700F/g at 1 to 15A/g, with 70% retention after 3000 cycles. When incorporated into an asymmetrical supercapacitor, $NiCo_2S_4$|reduced-graphene-oxide offered an energy-density of 31Wh/kg at a power-density of 987W/kg, and an energy-density of 23Wh/kg at a power-density of 7418W/kg. Nanoparticles of $NiCo_2S_4$ were embedded into nitrogen- and sulfur- co-doped graphene surfaces by using a 2-step hydrothermal process[360]. The material had a specific capacitance of 1420.2F/g at 10mV/s and 630.6F/g at 1A/g, with 76.6% retention at 10A/g. The energy-density was 19.35Wh/kg at a power-density of 235.0W/kg. Mesoporous hollow $NiCo_2S_4$ sub-microspheres and porous-graphene/single-walled-carbon nanotubes have been synthesized[361] without the use of templates. A resultant asymmetrical supercapacitor with graphene|single-walled-carbon-nanotubes and $NiCo_2S_4$ as the negative and positive electrodes, respectively, offered a maximum energy-density of 45.3Wh/kg at 800W/kg, with 87.5% retention after 20000 cycles.

NiMoS₄

Nanosheets of $NiMoS_4$ were anchored to reduced graphene oxide coated nickel foam by using a hydrothermal method[362]. The as-prepared $NiMoS_4$|reduced-graphene-oxide|Ni-foam had an hierarchical lamellar structure and had a specific capacity of 408mAh/g at 0.5A/g. When combined with an activated-carbon in a supercapacitor, the latter offered an energy-density of 67.4Wh/kg at 562.5W/kg.

NiS

Nanoporous crystalline nickel sulfide flakes could be hybridized[363] with reduced graphene oxide sheets. A composite which contained 40% of reduced graphene oxide had a specific capacitance of 1312F/g at a scan-rate of 5mV/s. This composite was used to construct an asymmetrical supercapacitor in conjunction with carbon as the negative electrode, giving a specific capacitance of 47.85F/g at 2A/g, an energy-density of 17.01Wh/kg and a power-density of 10kW/kg. Nickel-sulfide|N-doped-graphene nanocomposites were prepared[364] which had a specific capacitance of 1467.8F/g at 1A/g. An asymmetrical device which had the composite and doped-graphene as the electrodes, with polyurethane foam as a separator soaked with 6M KOH and graphite sheet as the current collector, offered an energy-density of 66.6Wh/kg at a power-density of 405.83W/kg, with 86.6% capacitance after 5000 cycles.

Ni_3S_4

A composite of Ni_3S_2 nanoparticles and 3-dimensional graphene was used[365] as a cathode in supercapacitors. The electrode capacitance increased by up to 111% when the composite was activated by voltammetric scanning in 1M KOH. The capacitance and diffusion coefficient of electrolyte ions in the activated composite electrode were about 3.7 and 6.5 times higher, respectively, than those of the Ni_3S_2 electrode. The activated composite electrode had a specific capacitance of 3296F/g at a current density of 16A/g. A composite of Fe_3O_4 nanoparticles and chemically reduced graphene oxide was meanwhile prepared as an anode. The Fe_3O_4|reduced-graphene-oxide electrode had a specific capacitance of 661F/g at 1A/g. An asymmetrical supercapacitor was constructed, by using the composite electrodes, which operated reversibly between 0 and 1.6V. It had a specific capacitance of 233F/g at 5mV/s, and offered a maximum energy-density of 82.5Wh/kg at a power-density of 930W/kg. Three-dimensional reduced-graphene-oxide wrapped Ni_3S_2 nanoparticles on nickel foam having a porous structure was synthesized[366] solvothermally. The Ni-foam|reduced-graphene-oxide|Ni_3S_2 composite had a specific capacitance of 4048mF/cm^2 (816.8F/g) at a current density of 5mA/cm^2 (0.98A/g); with 93.8% capacitance retention after 6000 cycles at a current density of 25mA/cm^2. An aqueous asymmetrical supercapacitor was constructed by using the Ni-foam|reduced-graphene-oxide|Ni_3S_2 as the positive electrode and N-doped graphene as the negative electrode. The device offered an energy-density of 32.6Wh/kg at a power-density of 399.8W/kg, and 16.7Wh/kg at 8000.2W/kg. A supercapacitor electrode consisted of vertically-aligned Ni_3S_2 mesoporous nanosheets on 3-dimensional reduced graphene oxide, supported by Ni foam[367]. The nanostructure had a specific capacitance of 1886F/g (1621F/g) at a current density of 1.0A/g (20.0A/g). A symmetrical supercapacitor was constructed by using these nanostructures, and offered an energy-density of 58.9Wh/kg, a power-density of 3.7kW/kg at 45.8Wh/kg and 92% capacitance retention after 30000 cycles at a constant current density of 10A/g.

Various sulfides (NiS, NiS_2, Ni_3S_2) were synthesized[368] on thermally reduced graphene oxide by using a 1-step hydrothermal process. Any Ni_xS_y formed microflowers with an average size of 1.1 to 2.3µm which grew homogeneously on the curved graphene nanosheets. The material had a specific capacitance of 1602.2F/g at 1A/g, with 90.81% retention after 3000 cycles at 5A/g. An asymmetrical supercapacitor was based on Ni_xS_y|graphene and graphene electrodes offered an energy-density of 39.78Wh/kg at a power-density of 0.75kW/kg; with 91.2% capacitance retention after 3000 cycles.

A Ni_3S_2|reduced-graphene-oxide composite was synthesized[369] *in situ*, in which the graphene oxide induced the oxidation of Ni^{2+} to Ni^{3+} and the morphology changed from microbar to polyhedral during hydrothermal processing. The resultant Ni_3S_2|reduced-graphene-oxide composite exhibited an energy storage capacity of 1830F/g at 2A/g. A resultant asymmetrical supercapacitor offered an energy-density of 37.3Wh/kg at a power-density of 398W/kg; with 91.4% capacitance retention after 10000 cycles at a current density of 2A/g. An electrode material was prepared[370] which had a composite structure of Ni_3S_2|NiV-layered-double-hydroxide-nanosheets on reduced-graphene-oxide|Ni-foam. The electrode had a specific capacitance of 3004.3F/g at a current density of 1A/g; with 92.2% capacitance retention after 5000 cycles. An asymmetrical supercapacitor which had this composite, and activated-carbon, as electrodes offered an energy-density of 59.4Wh/kg at a power-density of 852.3W/kg; with 98.3% capacitance retention after 8000 cycles. A graphene-encapsulated NiS|Ni_3S_2 nanostructure was synthesized[371] by using a polyethylene glycol-assisted 1-step hydrothermal method. The nanostructure had a specific capacity of 827C/g at 5A/g. A resultant asymmetrical supercapacitor offered a maximum energy-density of 86.3Wh/kg at 2A/g, with 98% retention after 5000 cycles.

SnS_2

Two-dimensional sulfide and graphene nanosheets were combined[372] to form a composite in which the sulfide nanosheets were uniformly distributed and enclosed within graphene layers. The composite had a specific capacitance of 565F/g, with 90% retention after 3000 cycles. A resultant symmetrical device offered an energy-density of 23.5Wh/kg and power-density 880W/kg at a current density of 1A/g.

SrS

The introduction of graphene oxide thin film improved the specific surface area and the electrical conductivity of strontium sulfide nanorods[373]. The resultant material had an energy-density of 1831.14F/g and 91.56Wh/kg at a current density of 3mA/cm². A symmetrical device offered an energy-density and power-density of 10.55Wh/kg and 294.35W/kg, respectively.

Yb_2S_3

Composite hydrophilic thin films of ytterbium sulfide, graphene oxide and graphene-oxide|Yb_2S_3 were synthesized[374] by using a binder-free ionic layer adsorption and reaction method. The Yb_2S_3, graphene oxide and graphene-oxide|Yb_2S_3 composite films had maximum specific capacitances of 181, 193 and 376F/g, respectively in 1M Na_2SO_4 electrolyte at a scan-rate of 5mV/s. A flexible symmetrical supercapacitor was constructed by using the graphene-oxide|Yb_2S_3 composite electrode as anode and cathode, and a flexible asymmetrical supercapacitor was constructed with graphene-oxide|Yb_2S_3 as the anode and MnO_2 as the cathode electrode, together with a polyvinylalcohol-Na_2SO_4 gel electrolyte. The symmetrical device had a specific capacitance of 58F/g, an energy-density of 23Wh/kg and a power-density of 0.43kW/kg. The asymmetrical device had a specific capacitance of 92F/g, an energy-density of 42Wh/kg and a power-density of 0.84kW/kg.

$(Zn,Co)S$

An electrode which comprised zinc cobalt sulfide nanosheets, supported on sandwich-like N-doped graphene|carbon-nanotube film, was created[375]. The $Zn_{0.76}Co_{0.24}S$ electrode had a specific capacitance of 2484F/g at 2A/g, with essentially no capacitance loss after 10000 cycles at 30A/g. An asymmetrical supercapacitor was constructed by using the composite as the positive electrode and N-doped graphene|carbon-nanotube film as the negative electrode. This device offered an energy-density of 50.2Wh/kg at 387.5W/kg, with 100% capacitance retention after 2000 cycles.

ZnS

Nanocomposites of ZnS-nanowires|analite(Cu_7S_4)-nanoparticles|reduced-graphene-oxide were prepared hydrothermally[376] and had a maximum specific capacitance of 1114F/g at a current-density of 1A/g; with 88% capacitance retention after 5000 cycles. This nanocomposite was used as the positive electrode and activated-carbon was used as the negative electrode in order to construct an asymmetrical supercapacitor device which offered a maximum energy-density of 22Wh/kg and a power-density of up to 595W/kg; with 77% capacitance retention after 5000 cycles. A nanoweb of ZnS was deposited directly onto nickel foam bearing a thin pre-deposited layer of hydrothermally prepared graphene oxide[377]. The conductivity of the graphene oxide supported ZnS nanoweb was 100.15S/cm and the specific surface area was 104.42m^2/g. The specific capacity was 3052F/g at a scan-rate of 2mV/s, and was 2400.30F/g at a current-density of 3mA/cm^2. The energy- and power-densities were 120Wh/kg at 3mA/cm^2 and 4407.73W/kg, respectively. A symmetrical device offered an energy-density of 20.29Wh/kg at $2mA/cm^2$.

Table 2. Summary of Ragone-plot data for supercapacitors with sulfide composite electrodes

Sulfide	Energy-Density (Wh/kg)	Power-Density (W/kg)
$CoMoS_2$	33.1	850
$CoMoS_2$	47.32	940
$CoNi_2S_4$	39.56	374.8
$CoNi_2S_4$	51.7	762
$CoNi_2S_4$	57.4	323
CoS	105.6	2500
CoS	50.3	415.8
CoS	36.6	800
CoS	10.56	2250
Co_3S_4	55.16	975
Co_3S_4	24.84	13000
Co_3S_4	54.32	6250
Co_3S_4	1.09	398
Co_3S_4	0.31	750
Co_9S_8	31.6	910
Co_9S_8	88.81	630
Co_9S_8	30.30	21800
MnS	14.9	66.5
MnS	12.8	4683.5
MoS_2	73.5	19800
MoS_2	265	18000
MoS_2	12.5	2500
MoS_2	36.43	400.0

MoS_2	1.12	8001.6
MoS_2	2.65	119.21
MoS_2	22.0	800
MoS_2	15	373
MoS_2	29.2	4517.7
MoS_2	26.6	125
MoS_2	26	500
MoS_2	17.8	1500
MoS_2	39.1	700
MoS_2	26.4	1015.4
MoS_2	87.38	683.94
$NiCo_2S_4$	21.9	417.1
$NiCo_2S_4$	13.5	2700
$NiCo_2S_4$	23.9	2460.6
$NiCo_2S_4$	20.9	800.2
$NiCo_2S_4$	30.4	800.0
$NiCo_2S_4$	25	54
$NiCo_2S_4$	46.7	1200.8
$NiCo_2S_4$	23	7418
$NiCo_2S_4$	19.35	235.0
$NiCo_2S_4$	45.3	800
$NiMoS_4$	67.4	562.5
NiS	17.01	10000
NiS	66.6	405.83
Ni_3S_4	82.5	930
Ni_3S_4	32.6	399.8
Ni_3S_4	16.7	8000.2

Ni₃S₄	37.3	398
Ni₃S₄	59.4	852.3
SnS₂	23.5	880
SrS	10.55	294.35
Yb₂S₃	23	430
Yb₂S₃	42	840
(Zn,Co)S	50.2	387.5
ZnS	22	595
ZnS	120	4407.73

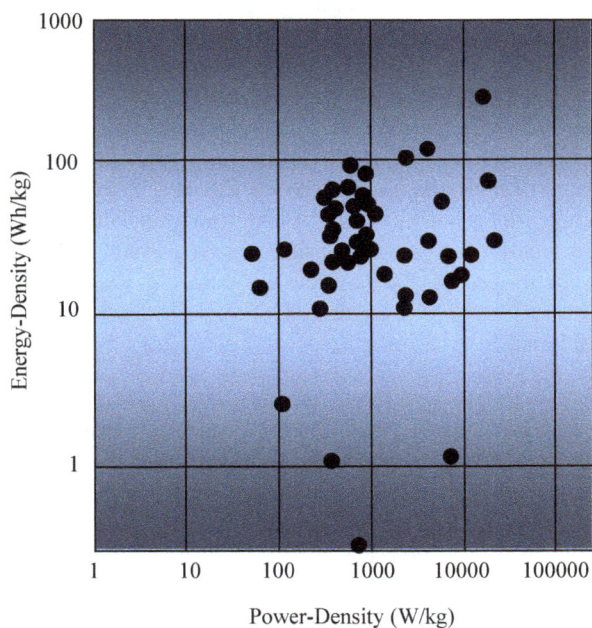

Figure 4. Ragone plot for sulfide composite electrodes

SELENIDES

CoSe₂

A CoSe₂ nanowire array on nickel foam served[378] as a positive electrode in an asymmetrical supercapacitor. The morphology of the CoSe₂ provided an improved electrical conductivity, numerous surface sites and short ion-diffusion paths. As-obtained CoSe₂ nanowire electrode material had an areal capacity of $1.08mAh/cm^2$ at $3mA/cm^2$. When combined with a Fe-TiN|N-doped-graphene negative electrode to construct an asymmetrical supercapacitor, the latter offered an energy-density of 91.8Wh/kg at a power-density of 281.4W/kg; with 94.6% capacitance retention after 10000 cycles.

NiCo₂Se₄

A NiCo₂Se₄|reduced-graphene-oxide electrode[379] had a specific capacitance of 1776F/g at a current density of 2A/g, with 51% capacitance retention at 50A/g. When this material was combined with sonochemically reduced graphene oxide to construct a supercapacitor, the device had a specific capacitance of 212F/g at 2A/g and offered a maximum energy-density of 66.2Wh/kg at a 1500W/kg power-density, with about 93.5% capacitance retention after 5000 cycles. A 1-step electrodeposition method was used[380] to grow NiCoSe₂|Ni₃Se₂ lamellar arrays directly onto N-doped graphene nanotubes, for use as free-standing positive electrodes for asymmetrical supercapacitors. As-constructed N-doped-graphene|NiCoSe₂|Ni₃Se₂ arrays had a specific capacitance of about 1308F/g at a current density of 1A/g, with about 1000F/g at 100A/g and essentially 100% capacitance retention after 10000 cycles. An asymmetrical supercapacitor which combined the above material with activated-carbon as the positive and negative electrodes respectively, offered an energy-density of 42.8Wh/kg at 2.6kW/kg, with some 94.4% capacitance retention after 10000 cycles.

VSe₂

A VSe₂|reduced-graphene-oxide composite was synthesized hydrothermally[381] which contained 0.15, 0.3 or 0.75wt% of graphene. A supercapacitor which included the 0.3wt% graphene composite had a specific capacitance of about 680F/g at 1A/g, an energy-density of about 212Wh/kg, a power-density of about 3.3kW/kg and with some 81% capacitance retention after 10000 cycles.

WSe₂

A nanosheet-like WSe²|reduced-graphene-oxide composite was prepared[382] hydrothermally which had a maximum specific capacitance of 389F/g at 1A/g, with 98.7% capacitance retention after 3000 cycles at a current-density of 7A/g. The energy-density was 34.5Wh/kg at 400W/kg, and 22.4Wh/kg at a power-density of 4000W/kg.

*Table 3. Summary of Ragone-plot data for supercapacitors
with selenide composite electrodes*

Selenide	Energy-Density (Wh/kg)	Power-Density (W/kg)
$CoSe_2$	91.8	281.4
$NiCo_2Se_4$	66.2	1500
$NiCo_2Se_4$	42.8	2600
VSe_2	212	3300
WSe_2	34.5	400
WSe_2	22.4	4000

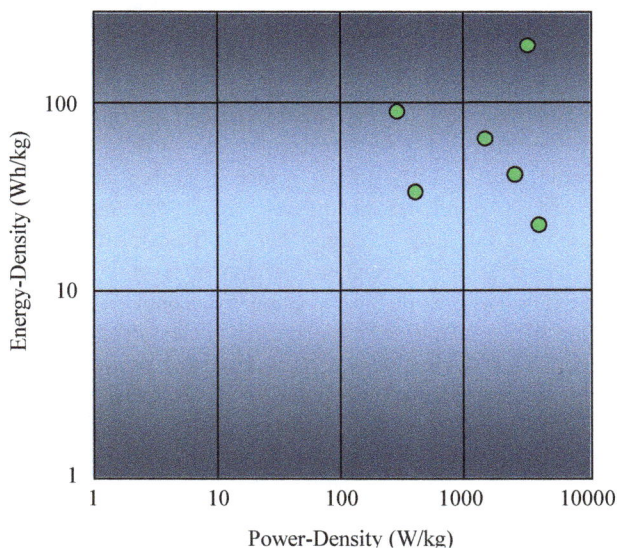

Figure 5. Ragone plot for selenide composite electrodes

HYDROXIDES

$Co(CO_3)(OH)$

Composites of reduced graphene oxide and cobalt carbonate hydroxide nanorods were prepared[383] which had a volumetric capacitance of $1627F/cm^3$ at a current-density of $0.5A/g$. The energy-density of a symmetrical supercapacitor which was constructed using the composite was $9.22mWh/cm^3$, with 100% retention after 10000 cycles at a current-density of $1A/g$.

$Co_5(CO_3)_2(OH)_6$

An asymmetrical supercapacitor was constructed[384] by using a composite of cobalt carbonate hydroxide nanowire-covered N-doped graphene as the positive electrode and porous N-doped graphene as the negative electrode. The composite had a specific capacitance of $1690F/g$ at $1.0A/g$, with 94.2% capacitance retention after 10000 cycles. The device had an areal capacitance of $153.5mF/cm^2$ at $1.0mA/cm^2$, and offered a power-density of $0.77Wh/m^2$ and $25.3W/m^2$.

$(Co,Ni)(OH)_2$

A nanocomposite of bimetallic hydroxide and polyaniline-modified partially reduced graphene oxide was created[385] for supercapacitor use. Large amounts of hydroxide nanosheet could grow longitudinally on the graphene surface substrate and form an hierarchical honeycomb-like micro/nanostructure array. The nanocomposite had a specific capacitance of $2760F/g$ at $1.0A/g$. After 1000 charge–discharge cycles, the nanocomposite exhibited 93.2% capacitance retention level at $10A/g$. A flexible asymmetrical supercapacitor device with polyvinylalcohol film electrolyte, and $CoNi(OH)_2$/graphene as the positive electrode, offered an energy-density of $74.84Wh/kg$ and power-density of $374.34W/kg$ at $0.5A/g$. Nanocomposite electrodes were prepared[386] from $Co_xNi_{1-x}(OH)_2$ and reduced graphene oxide, which consisted of porous hydroxide disks enclosed by graphene, and had a capacitance of 743 and $545C/g$ at 1 and $20A/g$, respectively. When combined with p-phenylenediamine-modified graphene, the device offered an energy-density of 72 and $44Wh/kg$ at power-densities of $797W/kg$ and $16.7kW/kg$, respectively.

$Co(OH)_2$

Nanocomposites of Co_3O_4|reduced-graphene-oxide have been prepared[387] by the co-precipitation of $Co(OH)_2$ and graphene oxide to form a precursor, followed by heat-treatment. A specific capacitance of $636F/g$ was measured when the mass ratio of Co_3O_4 to reduced graphene oxide was 80.3:19.7. A resultant asymmetrical supercapacitor, with the nanocomposite as the anode and activated-carbon as the cathode in 6M aqueous KOH solution as electrolyte, could cycle reversibly at a voltage of 0 to 1.5V and offer an energy-density of 35.7Wh/kg at a power-density of 225W/kg; with 95% capacitance retention after 1000 cycles at a current density of 0.625A/g. A composite of graphene and carbon nanotubes was synthesized[388] and coated with $Co(OH)_2$. Electrodes, one coated and one uncoated, were used in an asymmetrical supercapacitor which had a specific capacitance of 310F/g and offered an energy-density of 172Wh/kg and maximum power-density of 198kW/kg with an ionic liquid electrolyte. Nanosheets of $Co(OH)_2$ and amorphous FeOOH nanowires have been made[389] into transparent films, with the nanostructures encased in graphene shells. Thus extended the potential window of an FeOOH cathode from −0.8 to 0V to −1.2 to 0V. An asymmetrical transparent and flexible supercapacitor which was based upon $Co(OH)_2$|graphene and FeOOH|graphene electrodes had a specific capacitance of $25.5mF/cm^2$ and an energy-density of $1.04mWh/cm^3$; with 83.5% capacitance retention after 10000 cycles. An asymmetrical supercapacitor was constructed[390] by using graphene-supported $Co(OH)_2$ nanosheet as the positive electrode and carbon-fiber paper-supported activated-carbon as the negative electrode in KOH aqueous electrolyte. This offered an energy-density of 19.3Wh/kg at a power-density of 187.5W/kg, and a power-density of 3000W/kg at an energy-density of 16.7Wh/kg; with 100% capacitance retention after 20000 cycles. Cobalt hydroxide nanoparticles with a size of 50nm were deposited[391] onto reduced graphene oxide nanoflakes by using a hydrothermal method. The material was tested in 2M KOH by using a 3-electrode system. The electrode had a specific capacitance of 235.20F/g at 0.1A/g, with some 90% capacitance retention after 2000 cycles at 1.0A/g.

Porous flake-like α-$Co(OH)_2$ thin films have been prepared[392] by electrodeposition onto graphene nanosheets. Hydrophilic functional groups could act as anchoring sites and permit $Co(OH)_2$ to grow more easily on a functionalized graphene electrode. The density and thickness (13.1μm) of α-$Co(OH)_2$ deposits on the latter electrode was greater than that (12.3μm) on the plain electrode. The specific discharge capacitance of an α-$Co(OH)_2$|functionalized-graphene electrode decreased from $2149mF/cm^2$ to $1944mF/cm^2$ over 1000 cycles; 90%retention of the discharge capacitance. A resultant supercapacitor offered a power-density and energy-density of 1137W/kg and 43Wh/kg, respectively, at $8mA/cm^2$. Exfoliated α-$Co(OH)_2$ nanosheets having a high capacitance were arranged[393] on few-layer graphene. The hybrid material had a specific capacitance of 567.1F/g at 1A/g. When the hybrid nanocomposite was used as a positive electrode and activated-carbon was used as a negative electrode to construct an asymmetrical capacitor, the device offered an energy-density of 21.2Wh/kg at a power-density of 0.41kW/kg for a potential of 1.65V.

A composite of $Co(OH)_2$|reduced-graphene-oxide was synthesized[394] by using a 1-step cathodic electrodeposition method in a 2-electrode system at constant current density on a stainless-steel plate. A 1:4 weight-ratio of graphene-oxide to $CoCl_2 \bullet 6H_2O$ was the optimum, and the composite had a specific capacitance of 734F/g at a current density of 1A/g, with 95% capacitance retention after 1000 cycles. The average energy-density and power-density was 60.6Wh/kg and 3208W/kg, respectively. A series of α-$Co(OH)_2$/reduced-graphene-oxide microfilms was arranged[395] so as to improve the conductivity of α-$Co(OH)_2$ and prevent the re-stacking of α-$Co(OH)_2$ and reduced graphene oxide sheets. The optimum α-$Co(OH)_2$/reduced-graphene-oxide flexible electrode had a specific capacitance of $273.86mF/cm^2$ at $0.1mA/cm^2$. A resultant micro-supercapacitor offered a specific areal capacitance of $130F/cm^2$ at $0.5mA/cm^2$ and an energy-density of $20mWh/cm^3$ at a power-density of $56mW/cm^3$.

$Cu(OH)_2$

A 1-pot method used[396] to produce bundled nanorods of the hydroxide, embedded in a matrix of reduced graphene oxide. The material had a BET surface area of $78.7m^2/g$ and a distribution of structural pores and inter-particle pores. The composite had a capacitance of 602F/g at 0.2A/g in 1M KOH. A 2-electrode symmetrical device offered an energy-density and power-density of 84.5Wh/kg at 0.55kW/kg and 20.5Wh/kg at 5.5kW/kg.

$MnOOH$

A composite of manganese oxyhydroxide and 3-dimensional reduced graphene oxide was prepared[397] by using a 2-step hydrothermal method. Nanoneedles of the MnOOH grew on the porous 3-dimensional skeleton. The composite had a specific capacitance of 327F/g at 0.2A/g in 1M Na_2SO_4 electrolyte, with 96.7% capacitance retention after 1000 cycles. An asymmetrical supercapacitor was constructed by using the composite and activated-carbon as the electrodes. At a power-density of 378W/kg, the device could deliver a maximum energy-density of 52.7Wh/kg.

$Nd(OH)_3$

A neodymium hydroxide-nanorod|graphene composite was prepared[398] which had a specific capacitance of 820F/g at 1A/g, with 96% capacitance retention after 3000 cycles. An asymmetrical supercapacitor which comprised the composite, activated-carbon and 6M KOH electrolyte offered an energy-density of 40Wh/kg with 85.3% capacitance retention after 5000 cycles.

NiOH

A 1-pot hydrothermal method was used to grow vertically-oriented cobalt-nickel hydroxide and reduced graphene oxide nanosheets on conductive carbon cloth[399]. The as-prepared electrodes had a specific capacitance of 151.46F/g at 2.5A/g, with 88% retention after 1000 cycles. A symmetrical supercapacitor which comprised 2 composite electrodes offered an energy-density of 30.29Wh/kg at a power-density of 1500W/kg; with 85.6% retention after 3000 cycles.

Ni(OH)₂

Graphene-supported $Ni(OH)_2$ nanowires which were prepared[400] by using a hydrothermal method exhibited ultra-fast charge-discharge rates, and the capacitance arose mainly from battery-like behavior rather than pseudocapacitive behavior. The graphene-supported $Ni(OH)_2$ nanowires and carbon were used as positive and negative electrodes, respectively, in a supercapacitor. The latter offered a maximum specific power-density of 40840W/kg with an energy-density of about 17.3Wh/kg. A 1-step method was used[401] to prepare mechanically strong and electrically conductive graphene|$Ni(OH)_2$ composite hydrogels which had an interconnected porous network and could be used as 3-dimensional supercapacitor electrodes without adding a binder. An optimum composite which contained about 82wt%$Ni(OH)_2$ had a specific capacitance of about 1247F/g at a scan-rate of 5mV/s and about 785F/g at 40mV/s; with some 63% capacitance retention. The capacitance of these hydrogels was much greater than that, 309F/g at 40mV/s, of a physical mixture of graphene sheets and $Ni(OH)_2$ nanoplates. The same method was use to create graphene|carbon-nanotube|$Ni(OH)_2$ composite hydrogels which has a specific capacitance of about 1352F/g at 5mV/s and some 66% capacitance retention at 40mV/s. These composite hydrogels offered energy-densities of some 43 and 47Wh/kg, with power-densities of some 8 and 9kW/kg, respectively.

Homogenous $Ni(OH)_2$ deposits on graphene were obtained[402] by using an electrostatic induced stretch growth method which triggered changes in the morphology and ordered stacking of $Ni(OH)_2$ nanosheets on graphene and its crystallization. When as-prepared $Ni(OH)_2$|graphene composite was used in supercapacitors, the latter had a specific capacitance of 1503F/g at 2mV/s. A layered reduced-graphene-oxide|α-$Ni(OH)_2$ composite was produced[403] by using a non-hydrothermal route, with glucose as a templating agent for the growth of the layered hydroxide and a reducing agent for treating the graphene oxide. This led to a stacking of layered α-$Ni(OH)_2$ over reduced graphene oxide sheets. The specific capacitance was 1671.67F/g at a current density of 1A/g, with 81% capacitance retention after 2000 cycles. A resultant asymmetrical supercapacitor device was constructed by combining the reduced-graphene-oxide|α-$Ni(OH)_2$ with reduced graphene oxide, and offered an energy-density of 42.67Wh/kg at a power-density of 0.4kW/kg. Ultra-small $Ni(OH)_2$ nanoparticles were anchored to reduced graphene oxide sheets[404]. This led to a specific capacitance of 1717F/g at 0.5A/g. The nanoparticles were introduced into reduced graphene oxide sheets as a spacer to prevent stacking. The resultant sheets then had a capacitance of 182F/g at 100A/g. A resultant asymmetrical supercapacitor which was constructed from the two materials offered an energy-density of 75Wh/kg and a power-density of 40000W/kg.

Composites of the form, $Ni(OH)_2$|reduced-graphene-oxide, have been synthesized[405] by using a 2-step method in which $Ni(OH)_2$ deposition and graphene oxide reduction occurred separately. The product had a uniform porous lamellar structure with a high specific surface area. When used as a supercapacitor electrode, the composite had a specific capacitance of 2877F/g, with 86.5% capacitance retention after 4000 cycles, and an energy-density of 120.9Wh/kg at a power-density of 0.14kW/kg. Sulfonated graphene was produced[406] by modifying graphene oxide with the aryl diazonium salt of sulfanilic acid. A hydrogel composite was then formed by anchoring $Ni(OH)_2$ onto the sulfonated graphene by using hydrothermal methods. When the mass ratio of sulfonated graphene to $Ni(OH)_2$ was 1:5, the electrode material had a specific capacitance of 1817.5F/g at a current density of 1A/g. At a current density of 10A/g, the capacitance retention could be as high as 89.5% after 1000 cycles. A resultant flexible sulfonated-graphene and activated-carbon asymmetrical supercapacitor had a specific capacitance of 80.44F/g at 50mA/g. Nitrogen-doped graphene and Co–Ni layered double hydroxide was synthesized[407] by co-precipitation, leading to the formation of well-dispersed $(Co,Ni)(OH)_2$ nanoflakes anchored on the surface of nitrogen-doped graphene sheets. The specific capacitance was 2092F/g at a current density of $5mA/cm^2$, with 86.5% retention at a current density of 5 to $50mA/cm^2$. A resultant asymmetrical supercapacitor with the above composite as the positive electrode and activated-carbon as the negative electrode offered an energy-density of 49.4Wh/kg and a power-density of 101.97W/kg at a current density of $5mA/cm^2$.

A graphene-carbon nanotube aerogel was created[408] by direct cryo-desiccation from an aqueous dispersion of a graphene-oxide and carbon-nanotube hybrid, followed by high-temperature carbonization. The aerogel possessed a sheet-scroll conjoined structure and could act as a 3-dimensional template for the perpendicular immobilization of $Ni(OH)_2$ nanosheets. The capacitance of $Ni(OH)_2$|aerogel composite was 1208F/g at 1A/g, with 88% retention after 2000 cycles. An symmetrical supercapacitor which was based upon $Ni(OH)_2$|aerogel and aerogel electrodes offered an energy-density of 30Wh/kg at a power-density of 820W/kg. A polypyrrole|$Ni(OH)_2$|sulfonated-graphene-oxide ternary composite was prepared[409] via the step-wise loading of $Ni(OH)_2$ and polypyrrole into sulfonated graphene oxide using hydrothermal and oxidative polymerization processes. The ternary composite had a specific capacitance of 1632.5F/g at a current density of 1A/g in 6M KOH solution, with 86% capacitance retention after 1000 cycles. An asymmetrical supercapacitor which used the ternary composite as the cathode and activated-carbon as the anode had a specific capacitance of 224F/g at 1A/g, an energy-density of 79.6Wh/kg and a power-density of 0.8kW/kg; with 60% capacitance retention rate after 5000 cycles at 10A/g.

A ternary composite consisting of silver nanoparticles embedded in reduced graphene oxide and $Ni(OH)_2$ was synthesized[410] by using microwave-assisted reaction. The reduced graphene oxide supported the $Ni(OH)_2$ and prevented its re-stacking, while the anchored Ag nanoparticles acted as spacers and increased the surface area and the electrical conductivity. The Ag-graphene|$Ni(OH)_2$ composite had a specific capacitance of 1220F/g at 1A/g. A resultant asymmetrical supercapacitor which was based upon Ag-graphene|$Ni(OH)_2$ and activated-carbon electrodes had a maximum energy-density of 41.2Wh/kg at a power-density of 375W/kg. Free-standing exfoliated graphite was coated[411] with vertical arrays of $Ni(OH)_2$ and the resultant composite exhibited battery-type capacitive behavior. When combined with an activated-carbon anode, the resultant asymmetrical supercapacitor exhibited 84.5% capacitance retention after 20000 cycles at 8A/g, and offered an energy-density of 34.7Wh/kg at a power-density of 15kW/kg. A free-standing nitrogen-doped 3-dimensional rivet graphene film was densely coated with ultrafine $Ni(OH)_2$ nanoneedles[412]. The film had a high conductivity and high specific surface area, and the morphology of the $Ni(OH)_2$ nanoneedles exposed more active sites to the electrolyte. The $Ni(OH)_2$|N-doped-graphene electrode had a specific capacitance of 256.1mAh/g. A resultant asymmetrical supercapacitor offered an energy-density of 50.8Wh/kg at an average power-density of 452W/kg.

Hexagonal β-$Ni(OH)_2$ nanoplates were grown[413] onto graphene nanosheets by using a microwave hydrothermal method and a graphene|NiO|$Ni(OH)_2$ composite was then obtained by additional treatment involving NiO separation-out from the nanoplates. The graphene|$Ni(OH)_2$ composite had a specific capacitance of 1172F/g at a scan-rate of 5mV/s, and a specific capacitance of 1042F/g at a discharge current density of 3A/g. A graphene|NiO|$Ni(OH)_2$ composite had a rate capability of 684F/g at a discharge current-density of 24A/g, with about 92% capacitance retention at 3A/g. Yttrium-doped-$Ni(OH)_2$|graphene nanosheet heterostructures were prepared[414] by using a solvothermal method. The flower-like Y-$Ni(OH)_2$|graphene microspheres consisted of wrinkled nanoflakes and transparent graphene nanosheets offering an increased surface area and high electrical conductivity. The heterostructure had a specific capacity of 822.3C/g at 1A/g. A device which was based upon electrodes of the composite and of activated-carbon offered an energy-density of 81.6Wh/kg at 1129.8W/kg; with 83.28% retention after 7000 cycles.

Graphene-quantum-dot|Ni(OH)$_2$ composites on carbon cloth were prepared[415] hydrothermally and the material was used as an electrode which had a maximum specific capacitance of 1825F/g at a current density of 1A/g, with 83.5% capacitance retention after 8000 cycles. A resultant symmetrical supercapacitor, made by using the above material as positive and negative electrodes, offered an energy-density of 80.8Wh/kg at a power-density of 2021W/kg. Composite electrodes of the form, Ni(OH)$_2$|reduced-graphene-oxide, exhibit a good electrochemical performance in aqueous alkaline electrolytes. Replacement of the aqueous electrolyte with a lithium-ion gel polymer electrolyte led[416] to a supercapacitor having a wider operational window. The resultant Ni(OH)$_2$|reduced-graphene-oxide supercapacitor could operate at up to 3V, had a specific capacity of 85mAh/g and offered a specific energy of 36.1Wh/kg at a specific power of 77.01W/kg. Upon reducing the upper limit on potential to 2.5V, the supercapacitor had a specific capacitance of 6.7F/g, a specific energy of 12.4Wh/kg and a specific power of 6.8kW/kg. Spiral Ni(OH)$_2$ was supported on a free-standing layered film of N-doped-graphene|carbon-nanotube film[417]. A flexible film electrode of the material could act as a cathode having gravimetric and areal capacitances of 2130F/g at 2A/g and 2.88F/cm^2 at 1A/g; with essentially 100% capacitance retention after 30000 circles at a current density of 20A/g. An asymmetrical supercapacitor which had the above material, and activated-carbon, as electrodes offered an energy-density of 60Wh/kg at 800W/kg.

Table 4. Summary of Ragone-plot data for supercapacitors with hydroxide composite electrodes

Hydroxide	Energy-Density (Wh/kg)	Power-Density (W/kg)
(Co,Ni)(OH)$_2$	74.84	374.34
Co(OH)$_2$	35.7	225
Co(OH)$_2$	172	198000
Co(OH)$_2$	19.3	187.5
Co(OH)$_2$	16.7	3000
Co(OH)$_2$	43	1137
Co(OH)$_2$	21.2	410
Co(OH)$_2$	60.6	3208

Cu(OH)$_2$	84.5	550
Cu(OH)$_2$	20.5	5500
MnOOH	52.7	378
NiOH	30.29	1500
Ni(OH)$_2$	17.3	40840
Ni(OH)$_2$	43	8000
Ni(OH)$_2$	47	9000
Ni(OH)$_2$	42.67	400
Ni(OH)$_2$	75	40000
Ni(OH)$_2$	120.9	140
Ni(OH)$_2$	49.4	101.97
Ni(OH)$_2$	30	820
Ni(OH)$_2$	79.6	800
Ni(OH)$_2$	41.2	375
Ni(OH)$_2$	34.7	15000
Ni(OH)$_2$	50.8	452
Ni(OH)$_2$	81.6	1129.8
Ni(OH)$_2$	80.8	2021
Ni(OH)$_2$	12.4	6800
Ni(OH)$_2$	60	800

PHOSPHATES

(Co,Mn)$_2$P$_2$O$_7$

Ring-like $(Co_{0.55}Mn_{0.45})_2P_2O_7$ wrapped in nitrogen-doped graphene was prepared[418] by using a 1-step microwave technique. The circular pyrophosphate structures had a honeycomb-like form which aided faster ion and electron transport. The material had a specific capacity of 236.21mAh/g at 1A/g, with 86.7% capacitance retention after 10000 cycles. When combined with a biomass-derived carbon anode in an asymmetrical supercapacitor, the maximum energy-density of the device was 75.72Wh/kg at a power-density of 800W/kg, and 60.08Wh/kg at 8kW/kg.

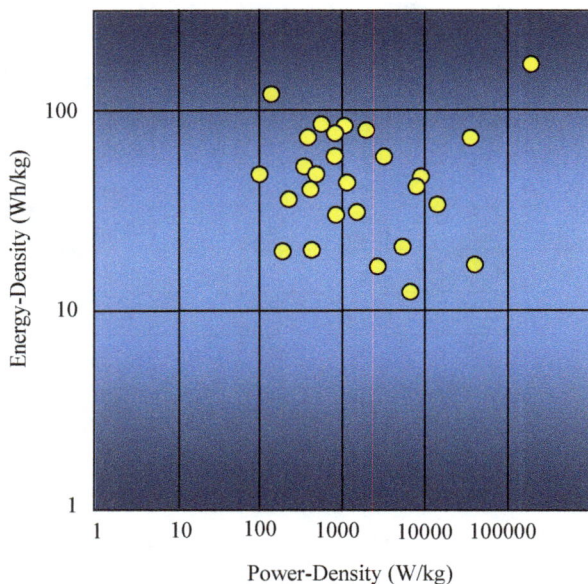

Figure 6. Ragone plot for hydroxide composite electrodes

K(Ni,Co)PO₄

A monolithic supercapacitor electrode consisted[419] of a $KNi_{0.67}Co_{0.33}PO_4 \bullet H_2O$|graphene composite hydrogel supported on nickel foam, and was prepared via a 1-step hydrothermal method which increased the surface area and mass-loading. This electrode had a capacitance of 3240mC/cm² (876C/g) at 2mA/cm²; with 78.3% retention at 100mA/cm². An asymmetrical supercapacitor was constructed by using $KNi_{0.67}Co_{0.33}PO_4 \bullet H_2O$|graphene|Ni-foam and Fe_2P|graphene-hydrogel|Ni-foam as the cathode and anode, respectively, and polyvinylalcohol-KOH as a gel electrolyte. It offered an energy-density of 69.2Wh/kg (3.9mWh/cm³) and a power-density of 13229W/kg (720mWh/cm³); 81.2% capacitance retention after 10000 cycles.

Materials Research Forum LLC

https://doi.org/10.21741/9781644901939

$Mn_3(PO_4)_2$

Nanosheets (2nm) of $Mn_3(PO_4)_2 \bullet 3H_2O$, prepared[420] by exfoliating bulk material in dimethylformamide under ultrasonication, spontaneously formed face-to-face stacks with exfoliated graphene. Resultant phosphate and graphene nanosheets with a mass ratio of 1:10 had a specific capacitance of 2086F/g at 1mV/s. When used to construct a supercapacitor with a polyvinylalcohol-KOH polymer electrolyte, the device had a specific capacitance of 152F/g (40mF/cm^2) at 0.5A/g, with an energy-density of 0.17μWh/cm^2 at 0.5A/g (1.3A/m^2) and a power-density of 46μW/cm^2 at 2A/g (5.3A/m^2); with essentially 100% capacitance retention after 2000 cycles at 2A/g.

$Na_5P_3O_{10}$

Polyphosphate-reduced-graphene-oxide|Ni-foam composite was synthesized[421] using various weight-ratios of $Na_5P_3O_{10}$ to graphene oxide. Composite with a weight-ratio of 2:1 had a specific capacitance of 118F/g at a current density of 5A/g, with 92% capacitance retention after 3000 cycles. When used to construct a symmetrical supercapacitor, the specific capacitance was 30.9F/g at 1A/g, the energy-density was 2.47Wh/kg at 523.48W/kg and the power-density was 2618W/kg at 0.8Wh/kg.

$NiCo(PO_4)_3$

Composites of the form, $NiCo(PO_4)_3$|40mg-graphene-foam, were synthesized[422] hydrothermally and used as a supercapacitor electrode with 1M KOH electrolyte. The latter composite had a specific capacity of 86.4mAh/g. When combined with an activated-carbon electrode, the resultant device offered a maximum energy-density of 34.8Wh/kg and a power-density of 377W/kg at a current of 0.5A/g; with 95% capacitance retention after 10000 cycles at a specific current of 8A/g.

$Ni_3(PO_4)_2$

The effect of differing contents of graphene foam upon the electrochemical capacitance of $Ni_3(PO_4)_2$ nanorods was examined[423], with regard to pristine $Ni_3(PO_4)_2$ nano-rods and $Ni_3(PO_4)_2$|graphene-foam composites with graphene contents of 30, 60, 90 or 120mg; plus 6M KOH electrolyte. A $Ni_3(PO_4)_2$|90mg-graphene-foam composite had the highest specific capacity: 48mAh/g at a current density of 0.5A/g. An asymmetrical device which was constructed with this composite as the cathode, and carbonized iron-cations adsorbed on polyaniline as the anode, in 6M KOH offered maximum energy and power densities of 49Wh/kg and 499W/kg, respectively, at 0.5A/g.

VOPO₄

A 3-dimensional vertically porous nanocomposite of layered vanadium phosphate and graphene nanosheets was prepared[424] which had a capacitance of 527.9F/g at a current density of 0.5A/g. Nanocomposite electrodes had a high surface area, and the structure provided short diffusion paths for electrolyte ions. An asymmetrical supercapacitor which was constructed by using vertically porous $VOPO_4$|graphene as the positive electrode and vertically porous 3-dimensional graphene as the negative electrode offered an energy-density of 108Wh/g. When used as a supercapacitor electrode, amorphous $VOPO_4$|graphene composite[425] had a specific capacitance of 508F/g at 0.5A/g and 359F/g at 10A/g, with 80% retention after 5000 cycles at 2A/g. It had an energy-density of 70.6Wh/kg with a power-density of 250W/kg.

Table 5. Summary of Ragone-plot data for supercapacitors
with phosphate composite electrodes

Phosphate	Energy-Density (Wh/kg)	Power-Density (W/kg)
$(Co,Mn)_2P_2O_7$	75.72	800
$(Co,Mn)_2P_2O_7$	60.08	8000
$K(Ni,Co)PO_4$	69.2	13229
$Na_5P_3O_{10}$	2.47	523.48
$Na_5P_3O_{10}$	0.8	2618
$VOPO_4$	70.6	250

NITRIDES

BN

A nanostructured hexagonal boron nitride and reduced graphene oxide composite was prepared[426] by inserting the h-BN into the graphene oxide via hydrothermal reaction. A specific capacitance of about 824F/g was measured at a current density of 4A/g. The potential window of a composite electrode ranged from -0.1 to 0.5V in 6M aqueous KOH electrolyte. This operating voltage increased to 1.4V in an asymmetrical supercapacitor when thermally reduced graphene oxide was used as the negative electrode and the composite was used as the positive electrode. The device had a specific capacitance of

145.7F/g at a current density of 6A/g and offered an energy-density of 39.6Wh/kg at a power-density of about 4200W/kg.

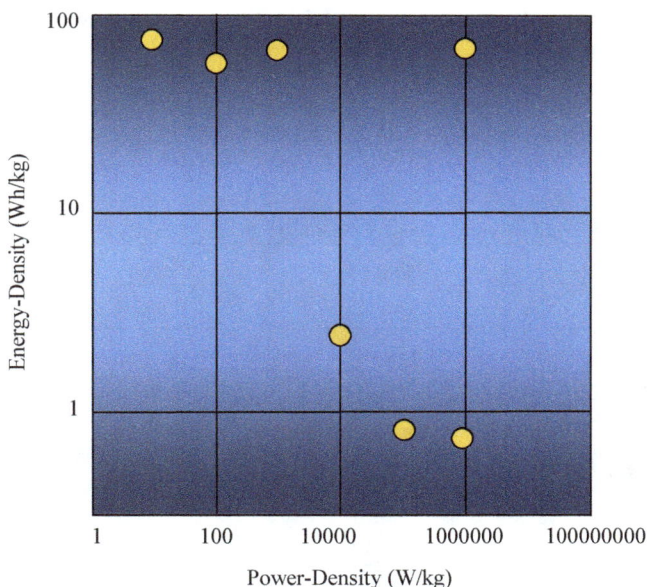

Figure 7. Ragone plot for phosphate composite electrodes

C_3N

Nitrogen-rich carbon derived from self-repairing graphitic C_3N_4 was combined[427] with graphene oxide to form a porous structure. The composite had a specific capacitance of 379.7F/g and an energy-density of 52.7Wh/kg at a current density of 0.25A/g, with 85% capacitance retention after 10000 cycles at a current density of 10A/g.

Ni_3N

A Ni_3N|graphene nanocomposite comprised[428] nitride nanoparticles anchored on reduced graphene oxide nanosheet. A 2-step oxidation/reduction reaction between Ni^I and Ni^{II} endowed the nanocomposite with a high capacitance; attaining a specific capacitance of 2087.5F/g at 1A/g. An asymmetrical supercapacitor which had ethylene glycol modified reduced graphene oxide as the negative electrode offered an energy-density of 50.5Wh/kg at 800W/kg.

VN

A composite of the form, VN|N-doped-graphene, was used[429] as the anode in a supercapacitor. The composite had a specific capacitance of 445F/g at 1A/g, with 98.66% capacitance retention after 10000 cycles at 10A/g. The electrode offered a maximum energy-density of about 81.73Wh/kg and a power-density of about 28.82kW/kg at 51.24Wh/kg.

Table 6. Summary of Ragone-plot data for supercapacitors
with nitride composite electrodes

Nitride	Energy-Density (Wh/kg)	Power-Density (W/kg)
BN	39.6	4200
Ni₃N	50.5	800
VN	81.73	28820

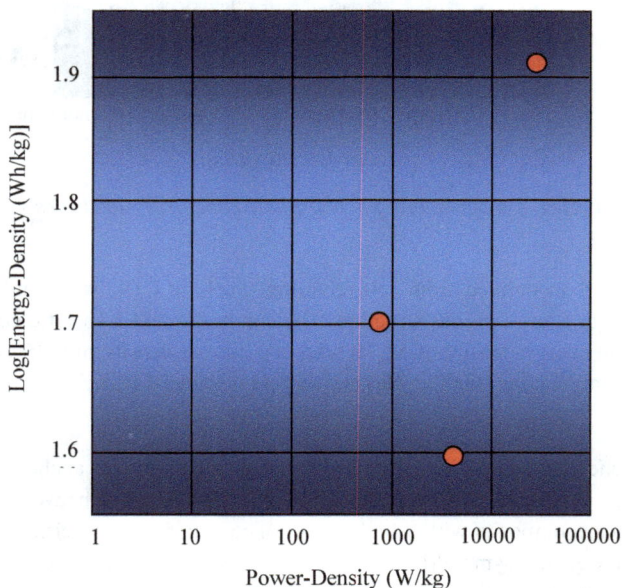

Figure 8. Ragone plot for nitride composite electrodes

MISCELLANEOUS

BiOBr

Graphene|BiOBr composites were prepared solvothermally[430] which had a specific capacity of 491C/g at 1A/g. An asymmetrical supercapacitor which had the composite as the negative electrode and Ni–Co–Al hydroxide as the positive electrode offered an energy-density of 29.2Wh/kg at a power-density of 700W/kg.

Cu_2O-$Cu(OH)_2$

An electrochemical deposition method was used[431] to prepare reduced-graphene-oxide|polypyrrole|Cu_2O-$Cu(OH)_2$ ternary nanocomposites. The Cu_2O-$Cu(OH)_2$ nanoparticles were dispersed on the surface of reduced-graphene-oxide|polypyrrole film with an average particle size of 50 to 70nm. In 0.5M Na_2SO_4 solution, the composite had a gravimetric specific capacitance of 997F/g at a current-density of 10A/g. There was a maximum energy-density of 20Wh/kg at a power-density of 8000W/kg and a maximum power-density of 1998.5W/kg at an energy-density of 5.8Wh/kg; with 90% capacitance retention after 2000 cycles.

Cu_3P

Hexagonal copper phosphide platelets were synthesized[432] by using chemical vapor deposition and combined with 3-dimensional graphene to produce a composite having a specific capacitance of 1095.85F/g at a 10mV/s scan-rate; with 95% capacitive retention after 3000 cycles at a current-density of 8.97A/g. An asymmetrical supercapacitor which had the composite as the cathode and activated-carbon as the anode had a specific capacity of108.78F/g, an energy-density of 8.23Wh/kg and a power-density of 439.6W/kg; with 96% retention after 5500 cycles.

Fe_3C

Porous carbon-coated carbide nanoparticles[433] have been loaded onto reduced graphene oxide nanosheets to form composites having a capacitance of 95.3mAh/g at 1A/g, with 81.5% retention after 5000 cycles. A device having a $Na_{0.5}MnO_2$ cathode and a composite anode plus 1M Na_2SO_4 or 6M KOH electrolytes offered energy-densities of 46.2Wh/kg at 1.2kW/kg and 28.3Wh/kg at 0.7kW/kg, respectively.

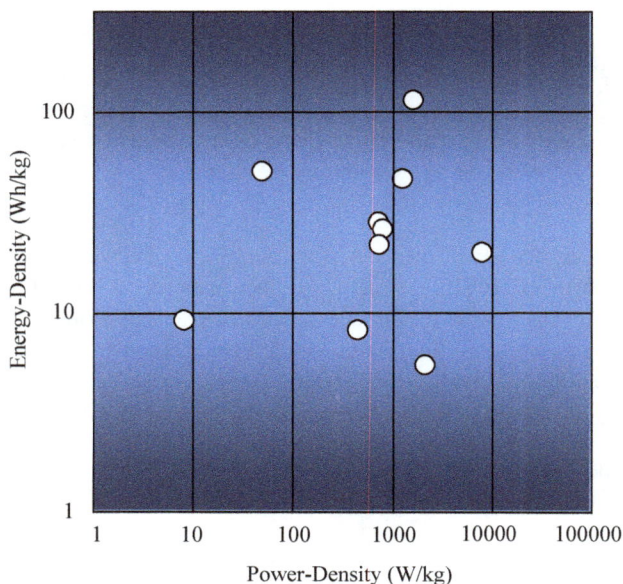

Figure 9. Ragone plot for miscellaneous composite electrodes

MnCO₃

A composite of rhombohedral $MnCO_3$ with reduced graphene oxide and multiwalled carbon nanotubes was synthesized[434] by using a hydrothermal method. The $MnCO_3$ had a specific capacitance of 185F/g at 1.5A/g in 1M Na_2SO_4 electrolyte. Maximum specific capacitances of 368F/g and 357F/g, respectively, were found for reduced-graphene-oxide|$MnCO_3$ and carbon-nanotube|$MnCO_3$ composites at 1.5A/g; with 96% and 97% capacitance retention, respectively, after 1000 cycles at 3A/g. They offered an energy-density of 51.11Wh/kg and 49.58Wh/kg, respectively, at a power delivery rate of 750W/kg. A 2V asymmetrical supercapacitor was constructed[435] by using reduced-graphene-oxide|carbon-nanofiber|$MnCO_3$ nanocomposite as the positive electrode and reduced graphene oxide as the negative electrode in a neutral 1M Na_2SO_4 aqueous electrolyte. Devices with symmetrical composite and reduced graphene oxide electrodes offered an energy-density of 4.8 and 3.6Wh/kg, respectively, at 0.1A/g. An asymmetrical device offered an energy-density of 21Wh/kg, with 97% capacitance retention after 1000 cycles at 1A/g. An energy-density of 15Wh/kg was retained when the power-density was increased to 1.07kW/kg. Rod-like $MnCO_3$ on reduced graphene oxide sheets was produced[436] by using a hydrothermal method. Chelating with citric acid aided the formation of a complex intermediate of Mn^{2+} and citrate ions and resulted in the formation of a 3-dimensional $MnCO_3$|reduced-graphene-oxide composite having an electrical conductivity of about 1056S/m, a surface area of $59m^2/g$ and a pore volume of $0.3cm^3/g$. The specific capacitance of the composite was about 1120F/g at a current density of 2A/g. An asymmetrical device which had the composite as the positive electrode and reduced graphene oxide as the negative electrode material had a specific capacitance of some 318F/g at 2A/g and offered an energy-density of about 113Wh/kg at 1600W/kg.

Table 7. Summary of Ragone-plot data for supercapacitors
with various composite electrodes

Material	Energy-Density (Wh/kg)	Power-Density (W/kg)
BiOBr	29.2	700
Cu_2O-$Cu(OH)_2$	20	8000
Cu_2O-$Cu(OH)_2$	5.8	1998.5
Cu_3P	8.23	439.6
Fe_3C	46.2	1200
Fe_3C	28.3	700
$MnCO_3$	51.11	49.58
$MnCO_3$	113	1600
Ni_2B	22.1	724.9
Ni_3Si_2	25.9	750

Ni_2B

A strongly-coupled nickel-boride|reduced-graphene-oxide composite was prepared[437] via the simultaneous chemical reduction of $NiCl_2\bullet6H_2O$ and graphene. The specific capacitance of the composite, when annealed at 200C, could attain 1073.4F/g at a current density of 1A/g in 6M KOH solution. A supercapacitor which had the composite as the positive electrode and activated-carbon as the negative electrode offered an energy-density of 22.1Wh/kg at a power-density of 724.9W/kg, with 72.4% capacitance retention after 2500 cycles at a current-density of 6A/g.

Ni$_3$Si$_2$

Nanostructures of Ni$_3$Si$_2$|NiOOH|graphene were synthesized[438] by means of low-pressure chemical vapor deposition. High-energy atoms in a carbon-rich atmosphere bombarded the nickel and silicon surfaces and catalyzed the growth of Ni-Si nanocrystals. The nanostructures had a specific capacity of 835.3C/g (1193.28F/g) at 1A/g. A resultant supercapacitor offered an energy-density of up to 25.9Wh/kg at 750W/kg.

Micro-Supercapacitors

When a graphene-quantum-dot||graphene-quantum-dot symmetrical micro-supercapacitor was prepared[439] by electrodeposition its electrochemical properties in aqueous and ionic-liquid electrolytes were such that the as-prepared micro-supercapacitor exhibited a good rate capability up to 1000V/s and an excellent power response, with very short relaxation time-constants: being 103.6µs in the aqueous electrolyte and 53.8µs in the ionic-liquid electrolyte. There was also an excellent cycling stability. When a graphene-quantum-dot||MnO$_2$ asymmetrical supercapacitor was constructed, with MnO$_2$ nanoneedles as the positive electrode and the dots as the negative electrode in an aqueous electrolyte, the specific capacitance and energy-density were both 2 times higher than those of the dot||dot symmetrical micro-supercapacitor.

Graphene-based in-plane interdigital micro-supercapacitors were prepared[440] on various substrates, showing that they could provide an areal capacitance of 80.7µF/cm^2 and a stack capacitance of 17.9F/cm^3. These were associated with a power-density of 495W/cm^3, making them superior to electrolytic capacitors, and an energy-density of 2.5mWh/cm^3, making them comparable to lithium thin-film batteries. There was again a good cycling stability, and they could operate at rates up to 1000V/s; three orders of magnitude greater than that of conventional supercapacitors. Ultrahigh-rate solid-state planar interdigital graphene-based micro-supercapacitors were prepared, via methane plasma-assisted reduction and photolithography of graphene oxide films on silicon wafers[441]. The electrochemical performance was markedly improved by increasing the number of interdigital fingers from 8 to 32, while reducing the finger-width from 1175 to 219µm. This led to an areal capacitance of 116µF/cm^2 and a stack-capacitance of 25.9F/cm^3, together with a power-density of 1270W/cm^3; again much higher than that of electrolytic capacitors. The energy-density was about 3.6mWh/cm^3; again comparable to that of lithium thin-film batteries. There was also a good cycling stability, with some 98.5% capacitance retention after 50kcycles. The device operated well at scan rates of up to 2000V/s; again 3 orders of magnitude higher than that of conventional supercapacitors.

It was proposed[442] that 2-dimensional nanochanneled graphene films of high packing density could be modelled on the vein of natural leaves. This would provide efficient ion transport pathways while maintaining a high energy-density. Such nanochannels would serve as pathways for efficient ion diffusion parallel to the graphene planes in micro-supercapacitors having an interdigitated electrode geometry. A simple solution-process was used[443] to fabricate micro-electrode patterns by choosing a water-dispersible graphene/sulfonated polyaniline material for the active micro-supercapacitor. A highly stabilized dispersion in aqueous solution permitted direct thin-film deposition onto flexible substrates and the formation of interdigital patterns by plasma-etching. The as-prepared solid-state micro-supercapacitors offered an areal capacitance of $3.31mF/cm^2$ and volumetric stack-capacitance of $16.55F/cm^3$; together with excellent rate and cycling performances. The micro-supercapacitors exhibited an energy-density of $1.51mWh/cm^3$, while maintaining a high power-density. The devices also exhibited good mechanical stability, and 96.5% of the capacitance was retained under various bending and twisting conditions. A one-step mask-assisted method for the fabrication of high-energy micro-supercapacitors was based[444] upon the interdigital hybrid electrode patterns of stacked phosphorene nanosheets and electrochemically exfoliated graphene in ionic-liquid electrolyte. Hybrid films with interdigital patterns were manufactured by the layer-by-layer deposition of phosphorene and graphene nanosheets, using an interdigital mask, and then transferred onto a flexible substrate. The resultant films were very uniform, flexible and conductive (319S/cm) and offered an energy-density of $11.6mWh/cm^3$; better than most nanocarbon-based micro-supercapacitors. There was only a slight capacitance fluctuation even when highly folded.

In the presence of carbon nanotubes of small diameter, laser-scribed graphene devices yielded[445] better energy storage. The dependence upon nanotube diameter was attributed to the effect of carbon nanotubes in preventing the re-stacking of laser-scribed graphene layers and thus increasing the ion-accessible surface area. Nanotubes having a smaller diameter could be more easily inserted between the scribed-graphene layers, thereby more effectively inhibiting re-stacking. The laser-scribed capacitor exhibited the best electrochemical properties when using single-wall nanotubes of 1 to 2nm diameter. These were a volumetric capacitance of $3.10F/cm^3$ at a current density of $1000mA/cm^3$, a volumetric energy-density of $0.84mWh/cm^3$ and a power-density of $1.0W/cm^3$; together with long-term cycling stability. Graphene/ZnO nanocomposite supercapacitor electrodes were prepared by laser-scribing[446]. A reduced graphene oxide and ZnO composite was created with a mass ratio of 1:25 of $Zn(NO_3)_2 \bullet 6H_2O$ to graphene oxide, and ZnO nanoparticles ranging from 20 to 50nm. There was a 12-fold times improvement in the specific capacitance at a current density of $0.1mA/cm^2$, as compared with pristine reduced graphene oxide electrodes. Flexible devices were prepared by spin-coating gel electrolyte, yielding a stack-capacitance of $9F/cm^3$ at a current-density of $150mA/cm^2$. The power-density was $70mW/cm^3$ and the energy-density was $1.2mWh/cm^3$.

A new type of fiber-based asymmetrical micro-supercapacitor, having two different graphene fiber-based electrodes, was constructed[447]. The electrochemical performances of $59.2mF/cm^2$ and $32.6mF/cm^2$ for both electrodes were promising. The potential window for such devices was widened to 1.6V, and the areal energy-density of $11.9\mu Wh/cm^2$ and volume energy-density of $11.9mWh/cm^3$ were the highest reported up to that time for such devices. There was also a good cycling stability, with an initial 92.7% capacitance retention after 8kcycles. In this connection, it was discovered[448] that a metal-needle spinneret was the key component in synthesizing neat graphene fibers having a porous surface via wet-spinning. As-prepared graphene fibers could possess a specific surface area of up to $839m^2/g$ combined with a specific capacitance of $228mF/cm^2$ at $39.7\mu A/cm^2$. Assembled micro-supercapacitors could exhibit energy-densities of $7.9\mu Wh/cm^2$ $(4.0mWh/cm^3)$; thus approaching those of lithium thin-film batteries. There was also long-term stability.

Vertically-oriented graphene nanosheets were synthesized[449] on highly-doped silicon substrates by using a simple method which was based upon cyclotron electron resonance plasma-enhanced chemical vapor deposition. The as-grown graphene electrodes were incorporated into a symmetrical micro-supercapacitor by using the aprotic ionic liquid. N-methyl-N-propylpyrrolidinium bis(trifluoromethylsulfonylimide) as an electrolyte. The resultant device offered a specific capacitance of $2mF/cm^2$, a power-density of $4mW/cm^2$ and an energy-density of $4\mu Wh/cm^2$ when operating at a cell voltage of 4V. The device retained 80% of the initial capacitance after 150000 galvanostatic charge-discharge cycles at a current-density of $1mA/cm^2$. The excellent electrochemical performance was attributed to the channel-based 3-dimensional graphene network, which facilitated rapid electrolyte ion-transport and short-range diffusion.

A heterostructure of graphene nanowall supported thin-layer nickel was used to form[450] an on-chip pseudocapacitor. This led to a specific energy-density of $2.1mWh/cm^3$ and power-density of up to $5.91W/cm^3$. These values were 2 orders of magnitude higher than those of electrolytic capacitors and thin-film batteries. The nickel was evaporated onto graphene nanowalls so as to serve as a shadow mask for micro-electrode patterning, and then as a precursor to *in situ* electrochemical conversion into pseudo-capacitive $Ni(OH)_2$. This led to the deposition of pseudo-capacitive material onto individual graphene nanoflakes, and exploited huge accessible surfaces.

Monolithic 3-dimensional graphene, as the electrode material, was used[451] to fabricate micro-supercapacitors via chemical vapor deposition. A seamless nickel catalyst substrate was first created by using polystyrene as a binder for nickel catalyst particles, followed by the sintering of individual nickel particles into a monolithic catalyst structure. The latter was penetrated by the carbon atoms which were generated by the decomposition of methane during chemical vapor deposition. The resultant wet 3-dimensional graphene was transferred onto a polyimide film. The final monolithic 3-dimensional graphene had a rough surface which comprised numerous graphene flakes. The oxygen-plasma exposed (30s) samples had an areal capacitance of $1.5mF/cm^2$ at a scan rate of 10V/s. A laser-processed graphene-based micro-planar supercapacitor exhibited[452] a volumetric energy-density which was 3.75 times that of existing micro-supercapacitors and 8785 times that of an aluminium electrolytic capacitor at an under 1000mV/s scan rate. It could moreover be tailored to various shapes, rolled up and plugged into interstitial spaces within a device. Such ultra-thin (18μm) micro-supercapacitor components could have a volumetric energy-density of $0.98mWh/cm^3$ in a LiCl-PVA gel and of $5.7mWh/cm^3$ in ionic liquid.

The scalable fabrication of graphene-based monolithic devices having various planar geometries was demonstrated[453] by the photochemical reduction of graphene-oxide/TiO_2 nanoparticle films. These micro-supercapacitors had a volumetric capacitance of $233.0F/cm^3$, together with marked flexibility and a great capacity for serial and parallel integration in aqueous gel electrolytes. By careful engineering of electrode|electrolyte interfaces, the monolithic devices could operate well in an ionic-liquid hydrophobic electrolyte (3.0V) at a scan rate of 200V/s; 2 orders of magnitude higher than those of conventional supercapacitors. The present supercapacitors exhibited a volumetric power-density of $312W/cm^3$ and an energy-density of $7.7mWh/cm^3$. These were perhaps the highest values reported thus far for carbon-based micro-supercapacitors.

Fiber-shaped micro-supercapacitors usually offer a low energy-density, because of inhomogeneity. A microfluidic method was used[454] to prepare homogeneous nitrogen-doped porous graphene fibers. The resultant capacitors used solid-state phosphoric acid/ polyvinyl alcohol and 1-ethyl-3-methylimidazolium tetrafluoroborate/poly(vinylidenefluoride-co-hexafluoropropylene) electrolytes which led to marked improvements in electrochemical performance. A capacitance of $1132mF/cm^2$, high cycling stability and long-term bending-durability were observed in the case of H_3PO_4/polyvinyl-alcohol electrolyte. Energy-densities of 95.7 to $46.9\mu Wh/cm^2$ and power-densities of 1.5 to $15W/cm^2$ were obtained in the case of the other electrolyte. The source of superior performance was deduced to be microfluidic-controlled fibers having a uniformly porous network, a high specific surface area of $388.6m^2/g$, optimum pyridinic nitrogen content (2.44%) and a conductivity of 30785S/m. A high-performance micro-supercapacitor was again based[455] upon microfluidic-oriented core-sheath polyaniline-nanorod-array/graphene composite fiber electrodes in which the polyaniline sheath array was wrapped *in situ* on the graphene core. The resultant device exhibited a capacitance of 230mF/cm^2, a conductivity of 18734S/m and an energy-density of 37.2μWh/cm^2, with 86.9% retention after 8000 cycles and long-term bending durability. The high performance was attributed to the core-sheath structure of the fibers. Their low energy-density and mechanical strength are nevertheless a problem for fiber-shaped supercapacitors. A new integration of carbon dots with graphene was used[456] to construct high-performance carbon-dot/graphene fiber-based devices, again using microfluidics. This led to an areal specific capacitance of 607mF/cm^2, a mass specific capacitance of 91.9F/g, long-term (2000 cycles) bending durability and an energy-density of 67.37μWh/cm^2. These properties were attributed to the large specific surface-area of 435.1m^2/g and ion channels with an average pore size of 2.5nm. A two-step electrochemical deposition method was used[457] to coat nickel fibers with reduced graphene oxide and MnO$_2$ so as to produce sheath-core flexible electrodes having an areal specific capacitance of 119.4mF/cm^2 at a current density of 0.5mA/cm^2 in 1M Na$_2$SO$_4$ electrolyte. Upon using polyvinyl alcohol plus LiCl as a solid-state electrolyte, and nickel reduced-graphene oxide and MnO$_2$ as flexible electrodes in a symmetrical fiber-shaped micro-supercapacitor device with a maximum areal capacitance of 26.9mF/cm^2, a power-density of 0.1W/cm^3 and an energy-density of up to 0.27mWh/cm^3 were observed. The device also exhibited a good flexibility and high cyclic stability. In order to increase the energy-storage capacity of microscale fiber electrodes while maintaining their high power-density, mass loadings (up to 42.5wt%) of RuO$_2$ nanoparticles were applied[458] to nanocarbon-based microfibers which were composed largely of fully-reduced graphene oxide, with a lower number of single-walled carbon nanotubes as nanospacers. This involved the spatially confined hydrothermal assembly of a highly porous and 3-dimensionally interconnected carbon structure, the impregnation of wet carbon structures with aqueous Ru^{3+} ions and the anchoring of RuO$_2$ nanoparticles to the reduced graphene oxide surfaces. Devices which were assembled from these fibers had a specific volumetric capacitance of 199F/cm^3 at 2mV/s and an energy-density of 27.3mWh/cm^3; the highest values reported at that time. A ternary binder-free nanocomposite on a carbon fiber substrate was developed[459] for use in fiber-shaped micro-supercapacitors. Synergistic effects of the various components of the electrodes yielded a specific capacitance of 2.9F/cm^2 (194F/cm^3 and 550mF/cm) at a current-density of 5mA/cm^2; with 95% capacitance retention after 5000 cycles in 1M Na$_2$SO$_4$ electrolyte. A fiber-shaped asymmetrical micro-device which was based upon the hybrid electrode exhibited a maximum energy-density of 295μWh/cm^2 (19mWh/cm^3) and power-density of $14mW/cm^2$ ($930mW/cm^3$) within an operational voltage window of 0 to 2.0V in a solid-state Na_2SO_4 electrolyte.

Hybrid nanocomposites which comprise metal-oxide nanoparticles and 3-dimensional graphene combine the advantages of metal oxides and graphene. Flexible micro-supercapacitors were based[460] upon single-walled carbon nanotube-bridged laser-induced graphene fibers, decorated with MnO_2 nanoparticles. The carbon nanotubes were deposited onto the laser-treated graphene surfaces. Due to synergistic effects arising from the conductive nanotube-bridged graphene network and the MnO_2 nanoparticles, the resultant flexible micro-supercapacitors offered an areal capacitance of $156.94mF/cm^2$; some 8 times higher than that ($20mF/cm^2$) of laser-induced MnO_2-decorated graphene-fiber devices. The present devices also exhibited an areal energy-density of $21.8\mu Wh/cm^2$, with 90.5% capacitance retention after 1200 bending cycles.

Printing

Planar graphene-based linear tandem micro-supercapacitors can be fabricated on various substrates by printing[461]. Such a device, consisting of 10 micro-supercapacitors could offer an output of 8.0V. They can possess great flexibility, without obvious capacitance degradation, for various degrees of bending. The areal capacitance can also be modulated by incorporating polyaniline-based pseudocapacitive nanosheets into graphene electrodes, leading to a capacitance of $7.6mF/cm^2$. In order to improve the output voltage and energy-density, asymmetrical linear tandem micro-supercapacitors can be made by the printing of linear-patterned graphene, as negative electrodes, and MnO_2 nanosheets as positive electrodes. Three serially connected micro-supercapacitors can yield a voltage of 5.4V.

Nanohybrid paper electrodes have been based[462] upon the full inkjet-printing synthesis of free-standing graphene paper-supported 3-dimensional porous graphene hydrogel polyaniline nanocomposite. A symmetrical device which was based upon such electrodes and a gel electrolyte exhibited great mechanical flexibility and offered an energy-density of 24.02Wh/kg at a power-density of 400.33W/kg.

Asymmetrical micro-supercapacitors that are flexible, binder-free and current-collector free can be made with an interdigitated architecture on a flexible transparent substrate by using masks and spray-coating[463]. The electrode materials comprise $Ti_3C_2T_x$ and reduced graphene oxide as 2-dimensional layers. These contribute to fast ion diffusion in the interdigitated architecture. The device operates at a voltage of 1V, and retains 97% of the initial capacitance after ten thousand cycles; with an energy-density of $8.6mWh/cm^3$ at a power-density of $0.2W/cm^3$.

Interdigital graphene-based micro-supercapacitors have been prepared[464] by chemical vapor deposition and photolithographic micro-fabrication. The energy-density could be to $34.48mWh/cm^3$ because of the interdigital architecture and monolayer graphene electrode. The capacitance was maintained at 93.80% after 2500 charge/discharge cycles, and the micro-supercapacitors could be operated at an ultra-high scan-rate of up to $5000V/s$.

Graphene-based micro-supercapacitors have been prepared[465], on Xerox-paper substrates, which had a maximum volumetric capacitance of $29.6mF/cm^3$ (volume of entire device) at $6.5mA/cm^3$. A hybrid electrode, with redox-active potassium iodide at the graphene surface, exhibited a volumetric capacitance of $130mF/cm^3$. The maximum energy-density for such a device, in H_2SO_4 electrolyte, was estimated to be $0.026mWh/cm^3$.

Multilayered graphene-based devices have been made[466] by the direct laser writing of stacked chemical vapor-deposited graphene films. By combining it with the dry transfer of multilayered graphene films, laser-writing permits the very efficient preparation of large-area micro-supercapacitors possessing great flexibility and various planar geometries. The devices exhibited an energy-density of $23mWh/cm^3$ and a power-density of $1860W/cm^3$ in an ionogel electrolyte. They exhibited an alternating-current line-filtering performance in polyvinyl-alcohol/H_2SO_4 hydrogel electrolyte, as indicated by a phase angle of $-76.2°$ at $120Hz$ and a resistance–capacitance constant of $0.54ms$. This was attributed to efficient ion transport, combined with a very good electrical conductivity of the planar multilayered-graphene micro-electrodes. Multilayered graphene plus polyaniline hybrid micro-supercapacitors which were made by direct laser writing exhibited an optimized capacitance of $3.8mF/cm^2$ in polyvinyl-alcohol/H_2SO_4 hydrogel.

Nitrogen- and oxygen- co-doped graphene quantum dots offer the possibility ultra-small size combined with rich active sites, high hydrophilicity and easy assembly into conductive carbon films. Electrophoresis has been used to construct carbon-based micro-supercapacitors by choosing such dots as the initial electrode material[467]. The doped devices have volumetric capacitances of $325F/cm^3$ in H_2SO_4, due to their high pseudocapacitive activity, high loading density and great electrolyte wetability; attributed to the large number of doped nitrogen and oxygen functional groups. The volumetric energy-density is higher than that of thin-film lithium batteries. Graphene quantum dots can thus solve the problem of carbon materials that have a low specific surface area and few active sites; improving the performance of planar on-chip micro-supercapacitors by reducing the interdigital width[468]. Again co-doping with nitrogen and oxygen to produce the material for narrow electrodes and gaps, both $6\mu m$, led to a power-density of

28.1µW/cm^2 and an energy-density of 15.11µWh/cm^2. This was combined with a capacitance retention rate of up to 96.1% after 10000 cycles of charging and discharging.

Plasma-reduced nitrogen-doped graphene oxide, with a nitrogen content of 8.05%, and ultra-fine MoO_2 nanoparticles with a diameter of 5 to 10nm, have been used[469] as electrode materials for high-energy flexible solid-state devices. When these are combined with plasma-reduced nitrogen-doped graphene oxide plus MoO_2 composite films, integrated on substrates such as cloth, glass and polyethylene terephthalate and laser-cut into micro-electrodes having various planar geometries, the resultant devices exhibit a working voltage of 1.4V, an areal capacitance of 33.6mF/cm^2, a volumetric capacitance of 152.9F/cm^3 at 5mV/s and an energy-density of 38.1mWh/cm^3. The devices also maintain stability during bending through up to 180°, with no obvious capacitance degradation following 1000 bendings though 60°.

A solid-state micro-supercapacitor with a patterned graphene-flake/polyethylene di-oxythiophene composite electrode was prepared[470] by using pen lithography over an area of 0.38cm^2. The device, when combined with a polyvinylalcohol/H_2SO_4 gel electrolyte, exhibited a maximum operating potential window of 1.2V, a specific capacitance of 37.08mF/cm^2 and an energy-density of 6.4mWh/cm^2 plus 89% capacitance retention after 2500 cycles.

Wet-jet milling exfoliation of graphite has been used[471] to scale-up the production of graphene as a supercapacitor material. Aqueous alcohol-based graphene inks permit metal-free flexible micro-supercapacitors to be screen-printed. The latter devices exhibit an areal capacitance of up to 1.324mF/cm^2 (5.296mF/cm^2 for a single electrode); corresponding to a volumetric capacitance of 0.490F/cm^3 (1.961F/cm^3 for a single electrode). These supercapacitors can also operate up to a power-density of more than 20mW/cm^2 at an energy-density of 0.064µWh/cm^2. They maintain excellent stability after 10000 charge-discharge cycles, 100 bends to a radius of 1cm and folding by angles of up to 180°.

The inkjet printing technique has been used[472] for the stacking of reduced graphene oxide and MoO_3 in order to construct flexible solid-state micro-supercapacitors. Ammonium molybdate tetrahydrate/graphene-oxide aqueous inks were easily printed on polyamide films and converted into reduced-graphene-oxide/MoO_3 hybrids by heat-treatments in air. The water-based inks could be inkjet-printed so as to form 2-dimensional materials. Inkjet-printed symmetrical micro-supercapacitors with a polyvinyl-alcohol/H_2SO_4 gel electrolyte had a voltage window of 0 to 0.8V, a volumetric specific capacitance of 22.5F/cm^3 at 0.044A/cm^3; together with great flexibility and good cyclic stability due to the synergistic effect of the reduced graphene oxide and MoO_3. Inkjet-printed composite micro-supercapacitors offered a maximum energy-density of $2mWh/cm^3$ and a power-density of $0.018W/cm^3$, with 82% capacitance persisting after 10000 charge-discharge cycles.

Because of the behavior of 2-dimensonal materials during the drying of thin films of liquid dispersions, the inkjet-printing of passivated graphene micro-flakes could create micro-supercapacitors having a 3-dimensional porous microstructure[473]. Macroscale through-thickness pores allowed fast ion-transport and improved the behavior of the devices; even in the presence of solid-state electrolytes. During multiple-pass printing, the porous microstructure absorbed further ink and thus the creation of 3-dimensional micro-supercapacitors.

Laser-induced graphene supercapacitors have been made[474] from a polyimide substrate decorated with red-mud nanoparticles and a solid-state ionic liquid electrolyte mixture of polyvinylidene fluoride, 1-ethyl-3-methylimidazolium bis(trifluoromethylsulfonyl)imide and 1-ethyl-3-methylimidazolium tetrafluoroborate; red mud being the iron-oxide-rich waste from aluminium production. The resultant 2-electrode device, with an interdigitated planar form and inkjet-printed silver current-collectors, exhibited an energy-density of $0.018mWh/cm^2$ at a power-density of $0.66mW/cm^2$; with 81% capacitance being retained after 4000 cycles. It also exhibited a good resistance to bending and flexure.

A stamping method has been developed[475] for the rapid preparation of graphene-based planar micro-supercapacitors which combines stamps of the desired shape with high-conductivity graphene inks. Flexible micro-supercapacitors of controlled structure can be prepared on various substrates without requiring metal current-collectors, additives or polymer binders. The resultant interdigitated micro-supercapacitors exhibit an areal capacitance of up to $21.7mF/cm^2$ at a current of $0.5mA$ and a power-density of $6mW/cm^2$ at an energy-density of $5\mu Wh/cm^2$. These devices maintain their performance after 10000 charge–discharge cycles and 300 bending cycles.

Another graphene composite ink was formulated[476] for the screen-printing of micro-supercapacitors in which the ink consisted of interwoven 2-dimensional graphene and activated-carbon nanofillers that were delaminated by 1-step sand-milling turbulent flow exfoliation. The embedding of activated carbon nanoplatelets into graphene layers markedly enhanced the electrochemical performance of screen-printed micro-supercapacitors to give an areal capacitance of $12.5mF/cm^2$; about 20 times that of pure graphene. The maximum energy-density, maximum power-density and cyclability were $1.07\mu Wh/cm^2$, $0.004mW/cm^2$ and 88.1% retention 5000 cycles, respectively, with 91.8% capacitance being retained after washing for 1.5h.

An *in situ* surface engineering method has been used[477] to improve the performance of graphene oxide micro-supercapacitor chips: 3-dimensionally printed graphene oxide chips were treated with parole monomer in order to ensure the spontaneous selective anchoring of polypyrrole to the micro-electrodes without affecting the spaces between the digitized electrodes. Edge-welded interface-reinforced graphene scaffolds exhibited a specific capacitance which ranged from 13.6 to 128.4mF/cm^2.

Planar micro-supercapacitors have been based[478] upon a graphene network and nickel hydroxide nanoplates by means of chemical vapor deposition, mask-free patterning and spray-coating. Devices having the optimum interdigital width exhibited a capacitance of 0.34mF/cm^2 at 10mV/s. Devices of the form, graphene/NiOOH/Ni(OH)$_2$, exhibited a capacitance of 0.75mF/cm^2 at 5mV/s (7.54F/cm^3) and a maximum energy-density of 1.04mWh/cm^3; with 80% capacitance being retained after 3000 charge-discharge cycles.

Planar on-chip micro-supercapacitors with winding interdigitated microelectrodes were prepared[479] using liquid-air interfacial assembly and lithography. Micro-supercapacitors with a finger-width of 210μm and a gap of 70μm could charge and discharge within 84ms, due to the shortening of the ion path. They also offered an area capacitance of 90.39μF/cm^2 and a stack capacitance of 29.16F/cm^3. The energy-density and power-density at 1V/s were 4.05mWh/cm^3 and 145.79mW/cm^3, respectively; with 98.5% capacitance retention after 2000 cyclic voltammetry cycles. They operated well at a scan-rate of 5000V/s.

In-plane asymmetrical micro-supercapacitors have been constructed[480] by using nano-sandwiched metal hexacyanoferrate/graphene hybrid thin films with interdigitated patterns. The voltage output could attain 1.8V, with an areal capacitance of up to 19.84mF/cm^2 and an energy-density of 44.6mWh/cm^3.

Ultraflexible MnO$_2$/reduced-graphene-oxide films were created[481] by using layer-by-layer coating and laser engraving. Due to the conductive flexible graphene oxide films, reduced by HI, the films exhibited an areal capacitance of 31.5mF/cm^2 at 0.2mA/cm^2. There was no degradation of the capacitance for bend radii ranging from ∞ to 0cm, and 77.0% retention of the initial capacitance after 6000 cycles. The electrochemical performance could be increased simply by adding more layers: upon adding up to 5 layers, the films exhibited an areal capacitance of 144.3mF/cm^2 at 0.3mA/cm^2 and an energy-density of 13.9mWh/cm^3 at 34.7mW/cm^3.

Graphene-based highly-integrated micro-supercapacitors have been produced[482] by using a continuous centrifugal coating technique, with the resultant highly-conductive graphene films acting as both patterned micro-electrodes and metal-free current-collectors and interconnects. This imparted a high integrity, marked flexibility and a tailored voltage and capacitance. The strong centrifugal and shear forces which were generated by the coating process produced graphene films which exhibited high alignment, compactness and packing density. This led to a volumetric capacitance of about $31.8F/cm^3$ and a volumetric energy-density of about $2.8mWh/cm^3$.

One self-patterned stamping process involves[483] the creation of pre-designed patterns on parafilm-coated polyethylene terephthalate substrates in order to fabricate flexible planar micro-supercapacitors. The imprinted patterns are filled with graphene-oxide/carbon-aerogel/MnO_2 hybrid paste, as a binder and additive-free active material, followed by grapheme oxide reduction using nascent hydrogen. Unlike hydrazine-based reduction, nascent hydrogen imparts integrity and stability to the active material without any peeling-away from the substrate. The prepared micro-supercapacitors exhibit large capacitances and an energy-density of $8.7mF/cm^2$ ($43.66F/cm^3$) and $6mWh/cm^3$ for interdigital electrodes or $14.2mF/cm^2$ ($71.34F/cm^3$) and $9.9mWh/cm^3$ for fractal electrodes, respectively. The micro-supercapacitors retained 85% capacitance after 25000 cycles.

Micro-supercapacitors have been prepared by means of photolithography and liquid/air interface self-assembly[484]. A uniform large-area film with a thickness of 18nm which was obtained in this way had a finger-gap of 30μm, and the electrode-width to interdigital-gap ratio was 3:1. The power-density and energy-density of the device were $1.21W/cm^3$ and $2.93mWh/cm^3$. The time-constant was 0.38ms, and 87.2% of the capacitance was retained after 10000 cycles.

Energy-storage micro-supercapacitors were prepared[485] which had a metal current collector-free symmetrical graphene-based planar structure. Exposure of the graphene to blue-violet laser-exposure and air plasma treatment increased the electrical conductivity by reducing functional groups. The wetability and the incidence of active sites were tuned by the air plasma treatment, so as to create a slightly functional group at the graphene surface. The resultant reduced graphene oxide had a resistance as low as $27.2\Omega/sq.$; thus ensuring good electron conductivity for fast electron transfer during electrochemical reactions. Electrochemical measurements revealed an areal capacitance of up to $21.86mF/cm^2$, together with a power-density of $5mW/cm^2$ and an energy-density of $2.49μWh/cm^2$. There was 99% retention after 10000 cycles.

A high-loading ($21.1mg/cm^2$) graphene-based micro-supercapacitor was prepared[486] by using hydrothermal, pressing and laser-engraving methods. The symmetrical micro-supercapacitors could have an areal capacitance as high as $569.5mF/cm^2$, together with an energy-density of $79.1μWh/cm^2$; with 98.8% capacitance retention after 20000 cycles.

The capacitance retention after 2000 bending cycles, for bending angles ranging from 0 to 180°, was 98.4%.

Fabrication of Fe_3O_4 nanoparticle-anchored laser-induced graphene having an hierarchical porous structure was achieved[487] on a flexible substrate. The aggregated oxide nanoparticles (~24.08nm), containing mesopores, were self-deposited in one step onto macroporous laser-induced graphene scaffolds. The 3-dimensional structures gave rise to superhydrophilic and capillary effects, with a pumping capacity of up to 0.096μm in water. This resulted in a continuous good wetability between the electrodes and a water-based electrolyte. Because of a reversible H^+-ion de-intercalation reaction with the oxide nanoparticles, micro-supercapacitors which had graphene/Fe_3O_4 as the anode and laser-induced graphene as the cathode exhibited an areal capacitance of 719.28mF/cm². This was 100 times higher than that of laser-induced graphene micro-supercapacitors. The areal energy-density of 60.20μWh/cm² was better than that of most previous devices of that type.

Three-dimensional porous graphene frameworks with thicknesses of up to 320μm have been grown by laser induction, directly onto polyimide, by optimizing the thermal sensitivity of the polyimide so as to increase the laser penetration-depth[488]. Hierarchical pore structures were obtained meanwhile due to the rapid liberation of gases during laser radiation, thus facilitating fast ion transport. The as-prepared 3-dimensional graphene offered a specific capacitance of 132.2mF/cm² at 0.5mA/cm²; almost an order-of-magnitude higher than that of most other laser-induced graphenes. When pseudocapacitive polypyrrole was then introduced into the graphene framework, the specific capacitance could be as high as 2412.2mF/cm² at 0.5mA/cm². Flexible solid-state micro-supercapacitors could be constructed which had an energy-density of 134.4μWh/cm² at a power-density of 325μW/cm²; with 95.6% capacitance retention after 10000 cycles.

Flexible micro-supercapacitors can be made[489] by means of single-pulse laser photonic-reduction stamping. In this way 1000 spatially-shaped laser pulses can be generated within one second and more than 30000 micro-supercapacitors can be produced within 600s. The micro-supercapacitor and narrow gaps are some dozens of microns or 500nm in size, respectively. Given the 3-dimensional structure of laser-induced graphene-based electrodes, a single micro-supercapacitor can exhibit an energy-density of 0.23Wh/cm³, a time-constant of 0.01ms, a specific capacitance of 128mF/cm² and 426.7F/cm³; together with long-term cyclability.

All-in-one planar micro-supercapacitor arrays can be based[490] upon hybrid electrodes in which ultra-thin ZnP nanosheets are anchored on 3-dimensional laser-induced graphene foams, arranged in an island-bridge architecture. These hybrid electrodes, with their large specific surface areas, can exhibit high ionic and electrical conductivities, an areal capacitance of 1425F/g (7.125F/cm^2) at 1A/g, an energy-density of 245mWh/cm^2 and a power-density of 12.50mW/kg at 145mWh/cm^2; plus long-term stability. It is suggested that the improved capacitance may result from the increased electrical conductivity and number of adsorbed charged electrolyte ions on the pseudocapacitive non-layered ultra-thin ZnP nanosheets. All-in-one planar micro-supercapacitor can similarly be produced[491] by using ZnO nanosheets which are again anchored to porous and 3-dimensional laser-induced porous graphene foams as electrode materials. This leads to rapid charge-transfer and to many diffusion channels. This hybrid approach can simultaneously exploit double-layer capacitance and faradaic energy-storage mechanisms. The ZnO/graphene electrodes can exhibit a specific capacitance of 14.7F/cm^2, together with an energy-density of 10.0Wh/kg and a power-density of 0.5Wh/kg.

Flexible micro-supercapacitors have been produced[492] by creating direct heating patterns on graphene oxide film. They are based upon planar reduced graphene-oxide/graphene-oxide structures, with reduced graphene oxide as the electrode and graphene oxide as the separator. The graphene oxide area which is heated by a heating pen is thereby reduced, while unheated parts continue to be graphene oxide. Interdigitated devices which are made at 400C can exhibit an areal capacitance of 94.8mF/cm^2 at a current density of 0.25mA/cm^2 within a polyvinyl-alcohol/H_2SO_4 gel electrolyte. They also provide an energy-density of 10.7mWh/cm^2 at a power-density of 112.6mW/cm^2, together with good mechanical stability; with an essentially unchanged areal capacitance after bending through angles ranging from 0° to 180°.

The main flaw of thin-film micro-supercapacitors, their low energy-density, can be countered by using thicker electrodes. A 3-dimensional printing method has been used[493] to create devices having interdigitated exfoliated-graphene/carbon-nanotube/silver-nanowire electrodes, with nanowelding of the nanowire junctions playing a critical role. In order to improve the electrochemical performance of the exfoliated graphene, some 1.7at% of phosphorus can be incorporated into the carbon framework. The areal capacitance of such 3-dimensional printed micro-supercapacitors can be 21.6mF/cm^2 at a scan-rate of 0.01V/s, with the areal energy-density ranging from 0.5 to 2μWh/cm^2 at a maximum power-density of 2.5mW/cm^2.

The reduction of graphene oxide is clearly a common method for obtaining graphene, but can suffer from re-stacking of the graphene sheets. A 2-dimensional composite electrode, based upon electrochemically reduced graphene oxide and polydopamine has however been created[494] in which the polydopamine was used as a chemical insert to impede re-stacking. The process involves the electroreduction of graphene oxide, followed by the

Materials Research Forum LLC

https://doi.org/10.21741/9781644901939

electro-oxidation of dopamine - both in the same electrolyte - by switching between cathodic and anodic potentials. An optimum composite electrode had a relaxation time of 0.88s, plus capacitances of 178F/g and 297F/cm^3 at 10mV/s and an excellent cycling stability at 100 to 2000mV/s after 30000 cycles. The dopamine between the graphene sheets prevented re-stacking and aided species diffusion within the composite, leading to an energy-density of 8.6mWh/cm^3 at a power-density of 7.8W/cm^3.

About the Author

Dr. Fisher has wide knowledge and experience of the fields of engineering, metallurgy and solid-state physics, beginning with work at Rolls-Royce Aero Engines on turbine-blade research, related to the Concord supersonic passenger-aircraft project, which led to a BSc degree (1971) from the University of Wales. This was followed by theoretical and experimental work on the directional solidification of eutectic alloys having the ultimate aim of developing composite turbine blades. This work led to a doctoral degree (1978) from the Swiss Federal Institute of Technology (Lausanne). He then acted for many years as an editor of various academic journals, in particular *Defect and Diffusion Forum*. In recent years he has specialized in writing monographs which introduce readers to the most rapidly developing ideas in the fields of engineering, metallurgy and solid-state physics. He is co-author of the widely-cited student textbook, *Fundamentals of Solidification*. Google Scholar credits him with 8394 citations and a lifetime h-index of 15.

References

[1] Chmiola, J., Yushin, G., Gogotsi, Y., Portet, C., Simon, P., Taberna, P.L., Science, 313[5794] 2006, 1760- 1763. https://doi.org/10.1126/science.1132195

[2] Das, H.T., Saravanya, S., Elumalai, P., ChemistrySelect, 3[46] 2018, 13275-13283. https://doi.org/10.1002/slct.201803034

[3] Li, T., Ma, R., Xu, X., Sun, S., Lin, J., Microporous and Mesoporous Materials, 324, 2021, 111277. https://doi.org/10.1016/j.micromeso.2021.111277

[4] Nanaji, K., Upadhyayula, V., Rao, T.N., Anandan, S., ACS Sustainable Chemistry and Engineering, 7[2] 2019, 2516-2529. https://doi.org/10.1021/acssuschemeng.8b05419

[5] Gómez-Urbano, J.L., Moreno-Fernández, G., Granados-Moreno, M., Rojo, T., Carriazo, D., Batteries and Supercaps, 4[11] 2021, 1749-1756. https://doi.org/10.1002/batt.202100134

[6] Sun, X., Liu, X., Li, F., Applied Surface Science, 551, 2021, 149438. https://doi.org/10.1016/j.apsusc.2021.149438

[7] Karaman, C., Karaman, O., Atar, N., Yola, M.L., Physical Chemistry Chemical Physics, 23[22] 2021, 12807-12821. https://doi.org/10.1039/D1CP01726H

[8] Xiong, C., Li, B., Duan, C., Dai, L., Nie, S., Qin, C., Xu, Y., Ni, Y., Chemical Engineering Journal, 418, 2021, 129518. https://doi.org/10.1016/j.cej.2021.129518

[9] Sun, D., Yu, X., Ji, X., Sun, Z., Sun, D., Journal of Alloys and Compounds, 805, 2019, 327-337. https://doi.org/10.1016/j.jallcom.2019.06.375

[10] Jia, M.Y., Li, Y., Xu, L.S., Yao, C.L., Jin, X.J., Journal of Wood Chemistry and Technology, 38[6] 2018, 417-429. https://doi.org/10.1080/02773813.2018.1488871

[11] Ren, X., Yuan, Z., Lin, Z., Lv, X., Qin, C., Jiang, X., JOM, 73, 2021, 2021, 4091-4102. https://doi.org/10.1007/s11837-021-04957-8

[12] Xu, L., Wang, H., Gao, J., Jin, X., Journal of Alloys and Compounds, 809, 2019, 151802. https://doi.org/10.1016/j.jallcom.2019.151802

[13] Xiong, C., Zou, Y., Peng, Z., Zhong, W., Nanoscale, 11[15] 2019, 7304-7316. https://doi.org/10.1039/C9NR00659A

[14] Jung, S., Myung, Y., Kim, B.N., Kim, I.G., You, I.K., Kim, T., Scientific Reports, 8[1] 2018, 1915. https://doi.org/10.1038/s41598-018-20096-8

[15] Choi, J.H., Lee, C., Cho, S., Moon, G.D., Kim, B.S., Chang, H., Jang, H.D., Carbon, 132, 2018, 16-24. https://doi.org/10.1016/j.carbon.2018.01.105

[16] Xu, X., Zhang, X., Zhao, Y., Hu, Y., Journal of Materials Science: Materials in Electronics, 29[10] 2018, 8410-8420. https://doi.org/10.1007/s10854-018-8852-3

[17] Sankar, S., Lee, H., Jung, H., Kim, A., Ahmed, A.T.A., Inamdar, A.I., Kim, H., Lee, S., Im, H., Young Kim, D., New Journal of Chemistry, 41[22] 2017, 13792-13797. https://doi.org/10.1039/C7NJ03136J

[18] Liu, X., Zou, S., Liu, K., Lv, C., Wu, Z., Yin, Y., Liang, T., Xie, Z., Journal of Power Sources, 384, 2018, 214-222. https://doi.org/10.1016/j.jpowsour.2018.02.087

[19] Veeramani, V., Sivakumar, M., Chen, S.M., Madhu, R., Alamri, H.R., Alothman, Z.A., Hossain, M.S.A., Chen, C.K., Yamauchi, Y., Miyamoto, N., Wu, K.C.W., RSC Advances, 7[72] 2017, 45668-45675. https://doi.org/10.1039/C7RA07810B

[20] Ma, L., Liu, R., Niu, H., Xing, L., Liu, L., Huang, Y., ACS Applied Materials and Interfaces, 8[49] 2016, 33608-33618. https://doi.org/10.1021/acsami.6b11034

[21] Karuppannan, M., Kim, Y., Sung, Y.E., Kwon, O.J., Journal of Applied Electrochemistry, 49[1] 2019, 57-66. https://doi.org/10.1007/s10800-018-1276-1

[22] Purkait, T., Singh, G., Singh, M., Kumar, D., Dey, R.S., Scientific Reports, 7[1] 2017, 15239. https://doi.org/10.1038/s41598-017-15463-w

[23] De Adhikari, A., Oraon, R., Tiwari, S.K., Lee, J.H., Nayak, G.C., RSC Advances, 5[35] 2015, 27347-27355. https://doi.org/10.1039/C4RA16174B

[24] Nanaji, K., Sarada, B.V., Varadaraju, U.V., N Rao, T., Anandan, S., Renewable Energy, 172, 2021, 502-513. https://doi.org/10.1016/j.renene.2021.03.039

[25] Pandey, S., Karakoti, M., Surana, K., Dhapola, P.S., SanthiBhushan, B., Ganguly, S., Singh, P.K., Abbas, A., Srivastava, A., Sahoo, N.G., Scientific Reports, 11[1] 2021, 3916. https://doi.org/10.1038/s41598-021-83483-8

[26] Lian, Y.M., Utetiwabo, W., Zhou, Y., Huang, Z.H., Zhou, L., Muhammad, F., Chen, R.J., Yang, W., Journal of Colloid and Interface Science, 557, 2019, 55-64. https://doi.org/10.1016/j.jcis.2019.09.003

[27] Elessawy, N.A., El Nady, J., Wazeer, W., Kashyout, A.B., Scientific Reports, 9[1] 2019, 1129. https://doi.org/10.1038/s41598-018-37369-x

[28] Wang, Q., Qin, B., Zhang, X.H., Wang, Y., Jin, L., Cao, Q., Diamond and Related Materials, 118, 2021, 108530. https://doi.org/10.1016/j.diamond.2021.108530

[29] Kuang, W., Yang, H., Ying, C., Gong, B., Kong, J., Cheng, X., Bo, Z., Waste Disposal and Sustainable Energy, 3[1] 2021, 31-39. https://doi.org/10.1007/s42768-020-00068-3

[30] Moustafa, E., El Nady, J., Kashyout, A.E.H.B., Shoueir, K., El-Kemary, M., ACS Omega. 6[36] 2021, 23090-23099. https://doi.org/10.1021/acsomega.1c02277

[31] Zhang, M., Sun, Z., Zhang, T., Qin, B., Sui, D., Xie, Y., Ma, Y., Chen, Y., Journal of Materials Chemistry A, 5[41] 2017, 21757-21764. https://doi.org/10.1039/C7TA05457B

[32] Wang, Y., Shi, Z., Huang, Y., Ma, Y., Wang, C., Chen, M., Chen, Y., Journal of Physical Chemistry C, 113[30; 2009, 13103-13107. https://doi.org/10.1021/jp902214f

[33] Liu, C., Yu, Z., Neff, D., Zhamu, A., Jang, B.Z., Nano Letters, 10[12] 2010, 4863-4868. https://doi.org/10.1021/nl102661q

[34] Fu, C., Kuang, Y., Huang, Z., Wang, X., Yin, Y., Chen, J., Zhou, H., Journal of Solid State Electrochemistry, 15[11-12] 2011, 2581-2585. https://doi.org/10.1007/s10008-010-1248-9

[35] Peng, X.Y., Liu, X.X., Diamond, D., Lau, K.T., Carbon, 49[11] 2011, 3488-3496. https://doi.org/10.1016/j.carbon.2011.04.047

[36] Qin, J., Zhang, M., Rajendran, S., Zhang, X., Liu, R., Materials Chemistry and Physics, 212, 2018, 30-34. https://doi.org/10.1016/j.matchemphys.2018.01.040

[37] Azizi, E., Arjomandi, J., Salimi, A., Lee, J.Y., Polymer, 195, 2020, 122429. https://doi.org/10.1016/j.polymer.2020.122429

[38] Soam, A., Mahender, C., Kumar, R., Singh, M., Materials Research Express, 6[2] 2019, 025054. https://doi.org/10.1088/2053-1591/aaf125

[39] Wang, W., Xiao, Y., Li, X., Cheng, Q., Wang, G., Chemical Engineering Journal, 371, 2019, 327-336. https://doi.org/10.1016/j.cej.2019.04.048

[40] Deng, L., Liu, J., Ma, Z., Fan, G., Liu, Z.H., RSC Advances, 8[44] 2018, 24796-24804. https://doi.org/10.1039/C8RA04200D

[41] Yari, A., Heidari Fathabad, S., Journal of Materials Science - Materials in Electronics, 31[16] 2020, 13051-13062. https://doi.org/10.1007/s10854-020-03855-0

[42] Xie, A., Tao, F., Li, T., Wang, L., Chen, S., Luo, S., Yao, C., Electrochimica Acta, 261, 2018, 314-322. https://doi.org/10.1016/j.electacta.2017.12.165

[43] Mazloum-Ardakani, M., Mohammadian-Sarcheshmeh, H., Naderi, H., Farbod, F., Sabaghian, F., Journal of Energy Storage, 26, 2019, 100998. https://doi.org/10.1016/j.est.2019.100998

[44] Sankar, K.V., Selvan, R.K., Electrochimica Acta, 213, 2016, 469-481. https://doi.org/10.1016/j.electacta.2016.07.056

[45] Mousa, M.A., Khairy, M., Shehab, M., Journal of Solid State Electrochemistry, 21[4] 2017, 995-1005. https://doi.org/10.1007/s10008-016-3446-6

[46] Haghshenas, M., Mazloum-Ardakani, M., Tamaddon, F., Nasiri, A., International Journal of Hydrogen Energy, 46[5] 2021, 3984-3995. https://doi.org/10.1016/j.ijhydene.2020.10.253

[47] Ahuja, P., Sahu, V., Ujjain, S.K., Sharma, R.K., Singh, G., Electrochimica Acta, 146, 2014, 429-436. https://doi.org/10.1016/j.electacta.2014.09.039

[48] Xia, X., Lei, W., Hao, Q., Wang, W., Wang, X., Electrochimica Acta, 99, 2013, 253-261. https://doi.org/10.1016/j.electacta.2013.03.131

[49] Veerasubramani, G.K., Krishnamoorthy, K., Kim, S.J., RSC Advances, 5[21] 2015, 16319-16327. https://doi.org/10.1039/C4RA15070H

[50] Jayasubramaniyan, S., Balasundari, S., Naresh, N., Rayjada, P.A., Ghosh, S., Satyanarayana, N., Muralidharan, P., Journal of Alloys and Compounds, 778, 2019, 900-912. https://doi.org/10.1016/j.jallcom.2018.11.187

[51] Xuan, H., Li, H., Yang, J., Liang, X., Xie, Z., Han, P., Wu, Y., International Journal of Hydrogen Energy, 45[11] 2020, 6024-6035. https://doi.org/10.1016/j.ijhydene.2019.12.178

[52] Jiang, X., Li, Z., Lu, G., Hu, N., Ji, G., Liu, W., Guo, X., Wu, D., Liu, X., Xu, C., Electrochimica Acta, 358, 2020, 136857. https://doi.org/10.1016/j.electacta.2020.136857

[53] Wang, H., Holt, C.M.B., Li, Z., Tan, X., Amirkhiz, B.S., Xu, Z., Olsen, B.C., Stephenson, T., Mitlin, D., Nano Research, 5[9] 2012, 605-617. https://doi.org/10.1007/s12274-012-0246-x

[54] Xiang, C., Li, M., Zhi, M., Manivannan, A., Wu, N., Journal of Power Sources, 226, 2013, 65-70. https://doi.org/10.1016/j.jpowsour.2012.10.064

[55] Liao, Q., Li, N., Jin, S., Yang, G., Wang, C., ACS Nano, 9[5] 2015, 5310-5317. https://doi.org/10.1021/acsnano.5b00821

[56] Ujjain, S.K., Singh, G., Sharma, R.K., Electrochimica Acta, 169, 2015, 276-282.

https://doi.org/10.1016/j.electacta.2015.03.141

[57] Xie, L., Su, F., Xie, L., Li, X., Liu, Z., Kong, Q., Guo, X., Zhang, Y., Wan, L., Li, K., Lv, C., Chen, C., ChemSusChem, 8[17] 2015, 2917-2926. https://doi.org/10.1002/cssc.201500355

[58] Yan, H., Bai, J., Liao, M., He, Y., Liu, Q., Liu, J., Zhang, H., Li, Z., Wang, J., European Journal of Inorganic Chemistry, 2017[8] 2017, 1143-1152. https://doi.org/10.1002/ejic.201601202

[59] Bai, Y., Liu, M., Sun, J., Gao, L., Ionics, 22[4] 2016, 535-544. https://doi.org/10.1007/s11581-015-1576-y

[60] Naderi, H.R., Norouzi, P., Ganjali, M.R., Gholipour-Ranjbar, H., Journal of Materials Science: Materials in Electronics, 28[19] 2017, 14504-14514. https://doi.org/10.1007/s10854-017-7314-7

[61] Deng, X., Li, J., Zhu, S., He, F., He, C., Liu, E., Shi, C., Li, Q., Zhao, N., Journal of Alloys and Compounds, 693, 2017, 16-24. https://doi.org/10.1016/j.jallcom.2016.09.096

[62] Ramesh, S., Karuppasamy, K., Kim, H.S., Kim, H.S., Kim, J.H., Scientific Reports, 8[1] 2018, 16543. https://doi.org/10.1038/s41598-018-34905-7

[63] Gao, Z., Chen, C., Chang, J., Chen, L., Wu, D., Xu, F., Jiang, K., Electrochimica Acta, 260, 2018, 932-943. https://doi.org/10.1016/j.electacta.2017.12.070

[64] Liu, C., Gao, A., Yi, F., Shu, D., Yi, H., Zhou, X., Hao, J., He, C., Zhu, Z., Electrochimica Acta, 326, 2019, 134965. https://doi.org/10.1016/j.electacta.2019.134965

[65] Khalaj, M., Sedghi, A., Miankushki, H.N., Golkhatmi, S.Z., Energy, 188, 2019, 116088. https://doi.org/10.1016/j.energy.2019.116088

[66] Raj, S., Srivastava, S.K., Kar, P., Roy, P., Electrochimica Acta, 302, 2019, 327-337. https://doi.org/10.1016/j.electacta.2019.02.010

[67] Jia, W.L., Li, J., Lu, Z.J., Juan, Y.F., Jiang, Y.Q., Journal of Alloys and Compounds, 815, 2020, 152373. https://doi.org/10.1016/j.jallcom.2019.152373

[68] Li, S., Yang, K., Ye, P., Ma, K., Zhang, Z., Huang, Q., Applied Surface Science, 503, 2020, 144090. https://doi.org/10.1016/j.apsusc.2019.144090

[69] Ansarinejad, H., Shabani-Nooshabadi, M., Ghoreishi, S.M., Chemistry - An Asian Journal, 16[10] 2021, 1258-1270. https://doi.org/10.1002/asia.202100124

[70] Muthu, R.N., Tatiparti, S.S.V., Journal of Materials Engineering and Performance,

29[10] 2020, 6535-6549. https://doi.org/10.1007/s11665-020-05176-z

[71] Khalaj, M., Golkhatmi, S.Z., Sedghi, A., Diamond and Related Materials, 114, 2021, 108313. https://doi.org/10.1016/j.diamond.2021.108313

[72] Syedvali, P., Rajeshkhanna, G., Umeshbabu, E., Kiran, G.U., Rao, G.R., Justin, P., RSC Advances, 5[48] 2015, 38407-38416. https://doi.org/10.1039/C5RA03463A

[73] Wang, J., Yang, J., Huang, T., Yin, W., RSC Advances, 6[106] 2016, 103923-103929. https://doi.org/10.1039/C6RA21281F

[74] Wang, J., Huang, T., Yin, W., Energy Technology, 5[11] 2017, 2055-2064. https://doi.org/10.1002/ente.201700154

[75] Tian, Z., Wang, X., Li, B., Li, H., Wu, Y., Electrochimica Acta, 298, 2019, 321-329. https://doi.org/10.1016/j.electacta.2018.12.103

[76] Fang, C., Zhang, D., Electrochimica Acta, 401, 2022, 139491. https://doi.org/10.1016/j.electacta.2021.139491

[77] Zhang, C., Xie, A., Zhang, W., Chang, J., Liu, C., Gu, L., Duo, X., Pan, F., Luo, S., Journal of Energy Storage, 34, 2021, 102181. https://doi.org/10.1016/j.est.2020.102181

[78] Viswanathan, A., Shetty, A.N., Electrochimica Acta, 257, 2017, 483-493. https://doi.org/10.1016/j.electacta.2017.10.099

[79] Luan, V.H., Han, J.H., Kang, H.W., Lee, W., Composites B, 178, 2019, 107464. https://doi.org/10.1016/j.compositesb.2019.107464

[80] Zhai, M., Li, A., Hu, J., RSC Advances, 10[60] 2020, 36554-36561. https://doi.org/10.1039/D0RA06758J

[81] Golkhatmi, S.Z., Khalaj, M., Izadpanahi, A., Sedghi, A., Solid State Sciences, 106, 2020, 106336. https://doi.org/10.1016/j.solidstatesciences.2020.106336

[82] Zhang, H., Gao, Q., Yang, K., Tan, Y., Tian, W., Zhu, L., Li, Z., Yang, C., Journal of Materials Chemistry A, 3[44] 2015, 22005-22011. https://doi.org/10.1039/C5TA06668A

[83] Wang, H., Xu, Z., Yi, H., Wei, H., Guo, Z., Wang, X., Nano Energy, 7, 2014, 86-96. https://doi.org/10.1016/j.nanoen.2014.04.009

[84] Xie, S., Zhang, M., Liu, P., Wang, S., Liu, S., Feng, H., Zheng, H., Cheng, F., Materials Research Bulletin, 96, 2017, 413-418. https://doi.org/10.1016/j.materresbull.2017.04.002

[85] Huang, X., Zhang, H., Li, N., Nanotechnology, 28[7] 2017, 075402.
https://doi.org/10.1088/1361-6528/aa542a

[86] Mallick, S., Jana, P.P., Raj, C.R., ChemElectroChem, 5[17] 2018, 2348-2356.
https://doi.org/10.1002/celc.201800521

[87] Meng, J., Wang, Y., Xie, X., Quan, H., Ionics, 25[10] 2019, 4925-4933.
https://doi.org/10.1007/s11581-019-03061-x

[88] Jayashree, M., Parthibavarman, M., Prabhakaran, S., Ionics, 25[7] 2019, 3309-3319.
https://doi.org/10.1007/s11581-019-02859-z

[89] Tian, Y., Hu, X., Wang, Y., Li, C., Wu, X., ACS Sustainable Chemistry and
Engineering, 7[10] 2019, 9211-9219.
https://doi.org/10.1021/acssuschemeng.8b06857

[90] Gupta, A., Sardana, S., Dal, J., Lather, S., Maan, A.S., Tripathi, R., Punia, R., Singh,
K., Ohlan, A., ACS Applied Energy Materials, 3[7] 2020, 6434-6446.
https://doi.org/10.1021/acsaem.0c00684

[91] Ding, B., Wu, X., Journal of Alloys and Compounds, 842, 2020, 155838.
https://doi.org/10.1016/j.jallcom.2020.155838

[92] Vigneshwaran, J., Abraham, S., Muniyandi, B., Prasankumar, T., Li, J.T., Jose, S.,
Surfaces and Interfaces, 27, 2021, 101572.
https://doi.org/10.1016/j.surfin.2021.101572

[93] Thadathil, A., Ismail, Y.A., Periyat, P., RSC Advances, 11[57] 2021, 35828-35841.
https://doi.org/10.1039/D1RA04946A

[94] Liu, M., Nan, H., Hu, X., Zhang, W., Qiao, L., Zeng, Y., Tian, H., Journal of Alloys
and Compounds, 864, 2021, 158147.
https://doi.org/10.1016/j.jallcom.2020.158147

[95] Ghanbari, R., Shabestari, M.E., Kali, E.N., Hu, Y., Ghorbani, S.R., Journal of the
Electrochemical Society, 168[3] 2021, 030543. https://doi.org/10.1149/1945-
7111/abef60

[96] Qu, Q., Yang, S., Feng, X., Advanced Materials, 23[46] 2011, 5574-5580.
https://doi.org/10.1002/adma.201103042

[97] Zhang, F., Zhang, T., Yang, X., Zhang, L., Leng, K., Huang, Y., Chen, Y., Energy
and Environmental Science, 6[6] 2013, 1623-1632.
https://doi.org/10.1039/c3ee40509e

[98] Ganganboina, A.B., Chowdhury, A.D., Doong, R.A., Electrochimica Acta, 245,

2017, 912-923. https://doi.org/10.1016/j.electacta.2017.06.002

[99] Mondal, S., Rana, U., Malik, S., Journal of Physical Chemistry C, 121[14] 2017, 7573-7583. https://doi.org/10.1021/acs.jpcc.6b10978

[100] Lalwani, S., Sahu, V., Marichi, R.B., Singh, G., Sharma, R.K., Electrochimica Acta, 224, 2017, 517-526. https://doi.org/10.1016/j.electacta.2016.12.057

[101] Pal, S., Majumder, S., Dutta, S., Banerjee, S., Satpati, B., De, S., Journal of Physics D: Applied Physics, 51[37] 2018, 375501. https://doi.org/10.1088/1361-6463/aad5b3

[102] Sheng, S., Liu, W., Zhu, K., Cheng, K., Ye, K., Wang, G., Cao, D., Yan, J., Journal of Colloid and Interface Science, 536, 2019, 235-244. https://doi.org/10.1016/j.jcis.2018.10.060

[103] Xie, S., Dong, F., Li, J., ChemistrySelect, 4[1] 2019, 437-440. https://doi.org/10.1002/slct.201803223

[104] Ao, J., Miao, R., Li, J., Journal of Alloys and Compounds, 802, 2019, 355-363. https://doi.org/10.1016/j.jallcom.2019.06.203

[105] Barmi, A.A.M., Aghazadeh, M., Moosavian, M.A., Golikand, A.N., Journal of Materials Science: Materials in Electronics, 31[18] 2020, 15198-15217. https://doi.org/10.1007/s10854-020-04085-0

[106] Mandal, D., Routh, P., Mahato, A.K., Nandi, A.K., ChemElectroChem, 6[19] 2019, 5136-5148. https://doi.org/10.1002/celc.201901280

[107] Zhang, H., Zhong, R., Liu, R., Mo, F., Wang, Y., Wu, X., Journal of Electroanalytical Chemistry, 898, 2021, 115632. https://doi.org/10.1016/j.jelechem.2021.115632

[108] Kumari, S., Maity, S., Vannathan, A.A., Shee, D., Das, P.P., Mal, S.S., Ceramics International, 46[3] 2020, 3028-3035. https://doi.org/10.1016/j.ceramint.2019.10.002

[109] Lu, M., Cao, Y., Xue, Y., Qiu, W., ACS Omega, 6[42] 2021, 27994-28003. https://doi.org/10.1021/acsomega.1c03863

[110] Fan, Q., Yang, M., Meng, Q., Cao, B., Yu, Y., Journal of the Electrochemical Society, 163[8] 2016, A1736-A1742. https://doi.org/10.1149/2.1271608jes

[111] Qian, Y., Cai, X., Zhang, C., Jiang, H., Zhou, L., Li, B., Lai, L., Electrochimica Acta, 258, 2017, 1311-1319. https://doi.org/10.1016/j.electacta.2017.11.188

[112] Xing, L.L., Wu, X., Huang, K.J., Journal of Colloid and Interface Science, 529,

2018, 171-179. https://doi.org/10.1016/j.jcis.2018.06.007

[113] Gao, Y.P., Zhai, Z.B., Wang, Q.Q., Hou, Z.Q., Huang, K.J., Journal of Colloid and Interface Science, 539, 2019, 38-44. https://doi.org/10.1016/j.jcis.2018.12.045

[114] Zhang, W., Chen, W., Li, L., Pang, S., Fan, X.I.N., Optoelectronics and Advanced Materials, Rapid Communications, 13[7-8] 2019, 463-471.

[115] Sankar, K.V., Selvan, R.K., Journal of Power Sources, 275, 2015, 399-407. https://doi.org/10.1016/j.jpowsour.2014.10.183

[116] Cheng, Q., Tang, J., Ma, J., Zhang, H., Shinya, N., Qin, L.C., Carbon, 49[9] 2011, 2917-2925. https://doi.org/10.1016/j.carbon.2011.02.068

[117] Choi, B.G., Yang, M., Hong, W.H., Choi, J.W., Huh, Y.S., ACS Nano, 6[5] 2012, 4020-4028. https://doi.org/10.1021/nn3003345

[118] Gao, H., Xiao, F., Ching, C.B., Duan, H., ACS Applied Materials and Interfaces, 4[5] 2012, 2801-2810. https://doi.org/10.1021/am300455d

[119] Perret, P.G., Malenfant, P.R.L., Bock, C., MacDougall, B., Journal of the Electrochemical Society, 159[9] 2012, A1554-A1561. https://doi.org/10.1149/2.064208jes

[120] Chen, C.Y., Fan, C.Y., Lee, M.T., Chang, J.KJournal of Materials Chemistry, 22[16] 2012, 7697-7700. https://doi.org/10.1039/c2jm16707g

[121] You, B., Li, N., Zhu, H., Zhu, X., Yang, J., ChemSusChem, 6[3] 2013, 474-480. https://doi.org/10.1002/cssc.201200709

[122] Sawangphruk, M., Srimuk, P., Chiochan, P., Krittayavathananon, A., Luanwuthi, S., Limtrakul, J., Carbon, 60, 2013, 109-116. https://doi.org/10.1016/j.carbon.2013.03.062

[123] Deng, S.X., Sun, D., Wu, C.H., Wang, H., Liu, J.B., Sun, Y.X., Yan, H., Electrochimica Acta, 111, 2013, 707-712. https://doi.org/10.1016/j.electacta.2013.08.055

[124] Cao, J., Wang, Y., Zhou, Y., Ouyang, J.H., Jia, D., Guo, L., Journal of Electroanalytical Chemistry, 689, 2013, 201-206. https://doi.org/10.1016/j.jelechem.2012.10.024

[125] Ge, J., Yao, H.B., Hu, W., Yu, X.F., Yan, Y.X., Mao, L.B., Li, H.H., Li, S.S., Yu, S.H., Nano Energy, 2[4] 2013, 505-513. https://doi.org/10.1016/j.nanoen.2012.12.002

[126] Chen, H., Zhou, S., Wu, L., ACS Applied Materials and Interfaces, 6[11] 2014,

8621-8630. https://doi.org/10.1021/am5014375

[127] Khoh, W.H., Hong, J.D., Colloids and Surfaces A, 456[1] 2014, 26-34. https://doi.org/10.1016/j.colsurfa.2014.05.003

[128] Bai, Z., Li, H., Li, M., Li, C., Wang, X., Qu, C., Yang, B., International Journal of Hydrogen Energy, 40[46] 2015, 16306-16315. https://doi.org/10.1016/j.ijhydene.2015.09.065

[129] Rajagopal, R., Kamaludeen, B.A., Krishnan, R., Electrochimica Acta, 180, 2015, 53-63. https://doi.org/10.1016/j.electacta.2015.08.087

[130] Jayakumar, A., Yoon, Y.J., Wang, R., Lee, J.M., RSC Advances, 5[114] 2015, 94388-94396. https://doi.org/10.1039/C5RA16884H

[131] Chen, J., Wang, Y., Cao, J., Liu, Y., Ouyang, J.H., Jia, D., Zhou, Y., Electrochimica Acta, 182, 2015, 861-870. https://doi.org/10.1016/j.electacta.2015.10.015

[132] Hao, J., Zhong, Y., Liao, Y., Shu, D., Kang, Z., Zou, X., He, C., Guo, S., Electrochimica Acta, 167, 2015, 412-420. https://doi.org/10.1016/j.electacta.2015.03.098

[133] Li, Y., Cao, D., Wang, Y., Yang, S., Zhang, D., Ye, K., Cheng, K., Yin, J., Wang, G., Xu, Y., Journal of Power Sources, 279, 2015, 138-145. https://doi.org/10.1016/j.jpowsour.2014.12.153

[134] Qian, L., Lu, L., Colloids and Surfaces A, 465, 2015, 32-38. https://doi.org/10.1016/j.colsurfa.2014.10.043

[135] Ghasemi, S., Hosseinzadeh, R., Jafari, M., International Journal of Hydrogen Energy, 40[2] 2015, 1037-1046. https://doi.org/10.1016/j.ijhydene.2014.11.072

[136] Xiong, C., Li, T., Khan, M., Li, H., Zhao, T., RSC Advances, 5[104] 2015, 85613-85619. https://doi.org/10.1039/C5RA14411F

[137] Liu, Y., Cai, X., Luo, B., Yan, M., Jiang, J., Shi, W., Carbon, 107, 2016, 426-432. https://doi.org/10.1016/j.carbon.2016.06.025

[138] Zhang, L., DeArmond, D., Alvarez, N.T., Zhao, D., Wang, T., Hou, G., Malik, R., Heineman, W.R., Shanov, V., Journal of Materials Chemistry A, 4[5] 2016, 1876-1886. https://doi.org/10.1039/C5TA10031C

[139] Lee, D.G., Kim, B.H., Synthetic Metals, 219, 2016, 115-123. https://doi.org/10.1016/j.synthmet.2016.06.007

[140] Liu, Z., Chen, W., Fan, X., Yu, J., Zhao, Y., Chinese Journal of Chemistry, 34[8]

2016, 839-846. https://doi.org/10.1002/cjoc.201600217

[141] Song, N., Wu, Y., Wang, W., Xiao, D., Tan, H., Zhao, Y., Materials Research Bulletin, 111, 2019, 267-276. https://doi.org/10.1016/j.materresbull.2018.11.024

[142] Wu, Y., Liu, S., Zhao, K., He, Z., Yuan, H., Lv, K., Jia, G., Ionics, 22[7] 2016, 1185-1195. https://doi.org/10.1007/s11581-015-1625-6

[143] Liu, Y., Miao, X., Fang, J., Zhang, X., Chen, S., Li, W., Feng, W., Chen, Y., Wang, W., Zhang, Y., ACS Applied Materials and Interfaces, 8[8] 2016, 5251-5260. https://doi.org/10.1021/acsami.5b10649

[144] Huang, Y., Zhu, M., Meng, W., Fu, Y., Wang, Z., Huang, Y., Pei, Z., Zhi, C., RSC Advances, 5[43] 2015, 33981-33989. https://doi.org/10.1039/C5RA02868J

[145] Hao, J., Liao, Y., Zhong, Y., Shu, D., He, C., Guo, S., Huang, Y., Zhong, J., Hu, L., Carbon, 94, 2015, 879-887. https://doi.org/10.1016/j.carbon.2015.07.069

[146] Xiong, C., Li, T., Dang, A., Zhao, T., Li, H., Lv, H., Journal of Power Sources, 306, 2016, 602-610. https://doi.org/10.1016/j.jpowsour.2015.12.056

[147] Zhao, T., Guo, S., Ji, X., Zhao, Y., Wang, X., Cheng, Y., Meng, J., Fullerenes Nanotubes and Carbon Nanostructures, 25[6] 2017, 391-396. https://doi.org/10.1080/1536383X.2017.1324850

[148] Yang, X., Niu, H., Jiang, H., Wang, Q., Qu, F., Journal of Materials Chemistry A, 4[29] 2016, 11264-11275. https://doi.org/10.1039/C6TA03474H

[149] Li, Z., An, Y., Hu, Z., An, N., Zhang, Y., Guo, B., Zhang, Z., Yang, Y., Wu, H., Journal of Materials Chemistry A, 4[27] 2016, 10618-10626. https://doi.org/10.1039/C6TA03358J

[150] Zheng, Y., Pann, W., Zhengn, D., Sun, C., Journal of the Electrochemical Society, 163[6] 2016, D230-D238. https://doi.org/10.1149/2.0341606jes

[151] Xu, L., Jia, M., Li, Y., Jin, X., Zhang, F., Scientific Reports, 7[1] 2017, 12857. https://doi.org/10.1038/s41598-017-11267-0

[152] Wang, P., Sun, S., Wang, S., Zhang, Y., Zhang, G., Li, Y., Li, S., Zhou, C., Fang, S., Journal of Applied Electrochemistry, 47[12] 2017, 1293-1303. https://doi.org/10.1007/s10800-017-1122-x

[153] Shivakumara, S., Munichandraiah, N., Solid State Communications, 260, 2017, 34-39. https://doi.org/10.1016/j.ssc.2017.05.015

[154] Amutha, B., Subramani, K., Reddy, P.N., Sathish, M., ChemistrySelect, 2[33] 2017, 10754-10761. https://doi.org/10.1002/slct.201701979

[155] Moussa, M., Shi, G., Wu, H., Zhao, Z., Voelcker, N.H., Losic, D., Ma, J., Materials and Design, 125, 2017, 1-10. https://doi.org/10.1016/j.matdes.2017.03.075

[156] Moussa, M., El-Kady, M.F., Wang, H., Michimore, A., Zhou, Q., Xu, J., Majeswki, P., Ma, J., Nanotechnology, 26[7] 2015, 075702. https://doi.org/10.1088/0957-4484/26/7/075702

[157] Ghosh, K., Yue, C.Y., Sk, M.M., Jena, R.K., ACS Applied Materials and Interfaces, 9[18] 2017, 15350-15363. https://doi.org/10.1021/acsami.6b16406

[158] Qiu, X., Xu, D., Ma, L., Wang, Y., International Journal of Electrochemical Science, 12[3] 2017, 2173-2183. https://doi.org/10.20964/2017.03.07

[159] Ma, W., Chen, S., Yang, S., Chen, W., Weng, W., Cheng, Y., Zhu, M., Carbon, 113, 2017, 151-158. https://doi.org/10.1016/j.carbon.2016.11.051

[160] Jana, M., Saha, S., Samanta, P., Murmu, N.C., Kim, N.H., Kuila, T., Lee, J.H., Journal of Power Sources, 340, 2017, 380-392. https://doi.org/10.1016/j.jpowsour.2016.11.096

[161] Sun, S., Jiang, G., Liu, Y., Yu, B., Evariste, U., Journal of Electronic Materials, 47[10] 2018, 5993-5999. https://doi.org/10.1007/s11664-018-6499-8

[162] Ahuja, P., Ujjain, S.K., Kanojia, R., Applied Surface Science, 427, 2018, 102-111. https://doi.org/10.1016/j.apsusc.2017.08.028

[163] Meng, X., Lu, L., Sun, C., ACS Applied Materials and Interfaces, 10[19] 2018, 16474-16481. https://doi.org/10.1021/acsami.8b02354

[164] Chi, H.Z., Wu, Y.Q., Shen, Y.K., Zhang, C., Xiong, Q., Qin, H., Electrochimica Acta, 289, 2018, 158-167. https://doi.org/10.1016/j.electacta.2018.09.025

[165] Arul, N.S., Han, J.I., Chen, P.C., ChemElectroChem, 5[19] 2018, 2747-2757. https://doi.org/10.1002/celc.201800700

[166] Zhao, J., Li, Y., Huang, F., Zhang, H., Gong, J., Miao, C., Zhu, K., Cheng, K., Ye, K., Yan, J., Cao, D., Wang, G., Zhang, X., Journal of Electroanalytical Chemistry, 823, 2018, 474-481. https://doi.org/10.1016/j.jelechem.2018.06.042

[167] Patil, D.S., Pawar, S.A., Shin, J.C., Kim, H.J., Journal of the Korean Physical Society, 72[8] 2018, 952-958. https://doi.org/10.3938/jkps.72.952

[168] Tseng, L.H., Hsiao, C.H., Nguyen, D.D., Hsieh, P.Y., Lee, C.Y., Tai, N.H., Electrochimica Acta, 266, 2018, 284-292. https://doi.org/10.1016/j.electacta.2018.02.029

[169] Wang, L., Ouyang, Y., Jiao, X., Xia, X., Lei, W., Hao, Q., Chemical Engineering

Journal, 334, 2018, 1-9. https://doi.org/10.1016/j.cej.2017.10.005

[170] Tian, W., Cheng, D., Wang, S., Xiong, C., Yang, Q., Applied Surface Science, 495, 2019, 143589. https://doi.org/10.1016/j.apsusc.2019.143589

[171] Kazemi, S.H., Aghdam, S.A., Journal of Electronic Materials, 48[8] 2019, 5088-5098. https://doi.org/10.1007/s11664-019-07318-z

[172] Singu, B.S., Yoon, K.R., Journal of Alloys and Compounds, 770, 2019, 1189-1199. https://doi.org/10.1016/j.jallcom.2018.08.145

[173] Sadak, O., Wang, W., Guan, J., Sundramoorthy, A.K., Gunasekaran, S., ACS Applied Nano Materials, 2[12] 2019, 4386-4394. https://doi.org/10.1021/acsanm.9b00797

[174] Liu, J.Q., Zhao, W., Wen, G.L., Xu, J., Chen, X., Zhang, Q., Wang, Y., Zhang, Y., Wu, Y.C., Journal of Alloys and Compounds, 787, 2019, 309-317. https://doi.org/10.1016/j.jallcom.2019.02.090

[175] Jadhav, S., Kalubarme, R.S., Terashima, C., Kale, B.B., Godbole, V., Fujishima, A., Gosavi, S.W., Electrochimica Acta, 299, 2019, 34-44. https://doi.org/10.1016/j.electacta.2018.12.182

[176] Zhu, C., Dong, X., Mei, X., Gao, M., Wang, K., Zhao, D., Journal of Materials Science, 55[36] 2020, 17108-17119. https://doi.org/10.1007/s10853-020-05212-2

[177] Rani, J.R., Thangavel, R., Kim, M., Lee, Y.S., Jang, J.H., Nanomaterials, 10[10] 2020, 2049, 1-16. https://doi.org/10.3390/nano10102049

[178] Ali, G.A.M., Journal of Electronic Materials, 49[9] 2020, 5411-5421. https://doi.org/10.1007/s11664-020-08268-7

[179] Dong, S., Wang, Z., Wang, J., Yao, Y., Liu, H., Materials Express, 10[8] 2020, 1308-1316. https://doi.org/10.1166/mex.2020.1766

[180] Zhang, M., Yang, D., Li, J., Vacuum, 178, 2020, 109455. https://doi.org/10.1016/j.vacuum.2020.109455

[181] Jain, R., Wadekar, P.H., Khose, R.V., Pethsangave, D.A., Some, S., Journal of Materials Science: Materials in Electronics, 31[11] 2020, 8385-8393. https://doi.org/10.1007/s10854-020-03373-z

[182] Tarimo, D.J., Oyedotun, K.O., Mirghni, A.A., Sylla, N.F., Manyala, N., Electrochimica Acta, 353, 2020, 136498. https://doi.org/10.1016/j.electacta.2020.136498

[183] Lv, Q., Hao, H., Ge, M., Li, W., Journal of Alloys and Compounds, 819, 2020,

152970. https://doi.org/10.1016/j.jallcom.2019.152970

[184] Hu, Z., Ma, K., Tian, W., Wang, F., Zhang, H., He, J., Deng, K., Zhang, Y.X., Yue, H., Ji, J., Applied Surface Science, 508, 2020, 144777. https://doi.org/10.1016/j.apsusc.2019.144777

[185] Zhang, M., Zheng, H., Zhu, H., Xu, Z., Liu, R., Chen, J., Song, Q., Song, X., Wu, J., Zhang, C., Cui, H., Vacuum, 176, 2020, 109315. https://doi.org/10.1016/j.vacuum.2020.109315

[186] Patil, S.H., Gaikwad, A.P., Waghmode, B.J., Sathaye, S.D., Patil, K.R., New Journal of Chemistry, 44[17] 2020, 6853-6861. https://doi.org/10.1039/C9NJ05898B

[187] Mane, V.J., Malavekar, D.B., Ubale, S.B., Bulakhe, R.N., In, I., Lokhande, C.D., Electrochimica Acta, 335, 2020, 135613. https://doi.org/10.1016/j.electacta.2020.135613

[188] Gong, D., Tong, H., Xiao, J., Li, T., Liu, J., Wu, Y., Chen, X., Liu, J., Zhang, X., Ceramics International, 47[23] 2021, 33020-33027. https://doi.org/10.1016/j.ceramint.2021.08.202

[189] Xie, A., Wang, H., Zhu, Z., Zhang, W., Li, X., Wang, Q., Luo, S., Surfaces and Interfaces, 25, 2021, 101177. https://doi.org/10.1016/j.surfin.2021.101177

[190] Wang, D.Y., Liu, J.Q., Chen, X., Yan, J., Wu, Y.C., Nano, 16[7] 2021, 2150080. https://doi.org/10.1142/S1793292021500806

[191] He, M., Cao, L., Li, W., Chang, X., Ren, Z., Journal of Alloys and Compounds, 865, 2021, 158934. https://doi.org/10.1016/j.jallcom.2021.158934

[192] Jangu, S., Satpathy, B.K., Raju, M., Jacob, C., Pradhan, D., Dalton Transactions, 50[20] 2021, 6878-6888. https://doi.org/10.1039/D1DT00422K

[193] Poochai, C., Sriprachuabwong, C., Sodtipinta, J., Lohitkarn, J., Pasakon, P., Primpray, V., Maeboonruan, N., Lomas, T., Wisitsoraat, A., Tuantranont, A., Journal of Colloid and Interface Science, 583, 2021, 734-745. https://doi.org/10.1016/j.jcis.2020.09.045

[194] Han, H., Sial, Q.A., Kalanur, S.S., Seo, H., Ceramics International, 46[10] 2020, 15631-15637. https://doi.org/10.1016/j.ceramint.2020.03.111

[195] Feng, G., Jiangying, Q., Zongbin, Z., Quan, Z., Beibei, L., Jieshan, Q., Carbon, 80[1] 2014, 640-650.

[196] Zhang, H., Huang, Z., Li, Y., Chen, Y., Wang, W., Ye, Y., Deng, P., RSC

Advances, 5[56] 2015, 45061-45067. https://doi.org/10.1039/C5RA05946A

[197] Mandal, M., Ghosh, D., Chattopadhyay, K., Das, C.K., Journal of Electronic Materials, 45[7] 2016, 3491-3500. https://doi.org/10.1007/s11664-016-4493-6

[198] Chen, S., Wang, L., Huang, M., Kang, L., Lei, Z., Xu, H., Shi, F., Liu, Z.H., Electrochimica Acta, 242, 2017, 10-18. https://doi.org/10.1016/j.electacta.2017.05.013

[199] Anilkumar, K.M., A., Manoj, M., Jinisha, B., Pradeep, V.S., Jayalekshmi, S., Electrochimica Acta, 236, 2017, 424-433. https://doi.org/10.1016/j.electacta.2017.03.167

[200] Gao, M., Wu, X., Qiu, H., Zhang, Q., Huang, K., Feng, S., Yang, Y., Wang, T., Zhao, B., Liu, Z., RSC Advances, 8[37] 2018, 20661-20668. https://doi.org/10.1039/C8RA00092A

[201] Boddula, R., Bolagam, R., Srinivasan, P., Ionics, 24[5] 2018, 1467-1474. https://doi.org/10.1007/s11581-017-2300-x

[202] Xiao, Y., Yang, M., Zhang, J., Zhang, A.Q., Zhou, J., Su, D., Zhou, L., Wang, X., Materials Research Express, 6[12] 2019, 125559. https://doi.org/10.1088/2053-1591/ab5f99

[203] Liu, X., Zhong, M., Yuan, L., Yang, F., Fu, Z., Xu, X., Wang, C., Tang, Y., Chemistry Letters, 49[8] 2020, 986-990. https://doi.org/10.1246/cl.200262

[204] Fan, L., Zhang, Y., Guo, Z., Sun, B., Tian, D., Feng, Y., Zhang, N., Sun, K., Chemistry - A European Journal, 26[42] 2020, 9314-9318. https://doi.org/10.1002/chem.201903947

[205] Wu, H., He, D., Wang, Y., Materials Letters, 268, 2020, 127613. https://doi.org/10.1016/j.matlet.2020.127613

[206] Low, W.H., Lim, S.S., Siong, C.W., Chia, C.H., Khiew, P.S., Ceramics International, 47[7] 2021, 9560-9568. https://doi.org/10.1016/j.ceramint.2020.12.090

[207] Thangappan, R., Arivanandhan, M., Kalaiselvam, S., Jayavel, R., Hayakawa, Y., Journal of Inorganic and Organometallic Polymers and Materials, 28[1] 2018, 50-62. https://doi.org/10.1007/s10904-017-0699-1

[208] Saeed, G., Kumar, S., Kim, N.H., Lee, J.H., Chemical Engineering Journal, 352, 2018, 268-276. https://doi.org/10.1016/j.cej.2018.07.026

[209] Liu, H., Zhang, M., Song, Z., Ma, T., Huang, Z., Wang, A., Shao, S., Journal of

Alloys and Compounds, 881, 2021, 160660.
https://doi.org/10.1016/j.jallcom.2021.160660

[210] Xia, X., Hao, Q., Lei, W., Wang, W., Wang, H., Wang, X., Journal of Materials Chemistry, 22[17] 2012, 8314-8320. https://doi.org/10.1039/c2jm16216d

[211] Ma, G., Li, K., Li, Y., Gao, B., Ding, T., Zhong, Q., Su, J., Gong, L., Chen, J., Yuan, L., Hu, B., Zhou, J., Huo, K., ChemElectroChem, 3[9] 2016, 1360-1368. https://doi.org/10.1002/celc.201600181

[212] Zhang, X., Deng, S., Zeng, Y., Yu, M., Zhong, Y., Xia, X., Tong, Y., Lu, X., Advanced Functional Materials, 28[44] 2018, 1805618. https://doi.org/10.1002/adfm.201805618

[213] Ojha, V., Kato, K., Kabbani, M.A., Babu, G., Ajayan, P.M., ChemistrySelect, 4[3] 2019, 1098-1102. https://doi.org/10.1002/slct.201803667

[214] Wu, K.F., Fan, J.H., Wang, X.H., Wang, M.T., Xie, X.F., Fan, J.T., Chen, A.Y., Energy and Fuels, 35[15] 2021, 12546-12555. https://doi.org/10.1021/acs.energyfuels.1c00932

[215] Zhou, Y., Wen, L., Zhan, K., Yan, Y., Zhao, B., Ceramics International, 44[17] 2018, 21848-21854. https://doi.org/10.1016/j.ceramint.2018.08.292

[216] Zhu, Y., Huang, H., Li, G., Liang, X., Zhou, W., Guo, J., Wei, W., Tang, S., Electrochimica Acta, 248, 2017, 562-569. https://doi.org/10.1016/j.electacta.2017.07.158

[217] Yu, M., Chen, J., Liu, J., Li, S., Ma, Y., Zhang, J., An, J., Electrochimica Acta, 151, 2015, 99-108. https://doi.org/10.1016/j.electacta.2014.10.156

[218] Xu, Y., Wang, L., Cao, P., Cai, C., Fu, Y., Ma, X., Journal of Power Sources, 306, 2016, 742-752. https://doi.org/10.1016/j.jpowsour.2015.12.106

[219] Yue, S., Tong, H., Lu, L., Tang, W., Bai, W., Jin, F., Han, Q., He, J., Liu, J., Zhang, X., Journal of Materials Chemistry A, 5[2] 2017, 689-698. https://doi.org/10.1039/C6TA09128H

[220] Lv, Y., Wang, H., Xu, X., Shi, J., Liu, W., Wang, X., Chemical Engineering Journal, 326, 2017, 401-410. https://doi.org/10.1016/j.cej.2017.05.167

[221] Li, Q., Lu, C., Chen, C., Xie, L., Liu, Y., Li, Y., Kong, Q., Wang, H., Energy Storage Materials, 8, 2017, 59-67. https://doi.org/10.1016/j.ensm.2017.04.002

[222] Mao, J.W., He, C.H., Qi, J.Q., Zhang, A.B., Sui, Y.W., He, Y.Z., Meng, Q.K., Wei, F.X., Journal of Electronic Materials, 47[1] 2018, 512-520.

https://doi.org/10.1007/s11664-017-5809-x

[223] Wang, Y., Zhang, M., Li, Y., Ma, T., Liu, H., Pan, D., Wang, X., Wang, A., Electrochimica Acta, 290, 2018, 12-20. https://doi.org/10.1016/j.electacta.2018.09.060

[224] Feng, H., Gao, S., Shi, J., Zhang, L., Peng, Z., Cao, S., Electrochimica Acta, 299, 2019, 116-124. https://doi.org/10.1016/j.electacta.2018.12.177

[225] Jiu, H., Jiang, L., Gao, Y., Zhang, Q., Zhang, L., Ionics, 25[9] 2019, 4325-4331. https://doi.org/10.1007/s11581-019-02970-1

[226] Chen, Y., Liu, T., Zhang, L., Yu, J., Applied Surface Science, 484, 2019, 135-143. https://doi.org/10.1016/j.apsusc.2019.04.074

[227] Jin, F., Tong, H., Lu, L., Meng, Q., Yue, S., Ding, B., Zhang, X., Journal of Alloys and Compounds, 787, 2019, 36-44. https://doi.org/10.1016/j.jallcom.2019.02.011

[228] Tong, H., Yue, S., Jin, F., Lu, L., Meng, Q., Zhang, X., Ceramics International, 44[3] 2018, 3113-3121. https://doi.org/10.1016/j.ceramint.2017.11.078

[229] Sethi, M., Shenoy, U.S., Bhat, D.K., New Journal of Chemistry, 44[10] 2020, 4033-4041. https://doi.org/10.1039/C9NJ05725K

[230] Patil, A.M., Yue, X., Yoshida, A., Li, S., Hao, X., Abudula, A., Guan, G., Applied Materials Today, 19, 2020, 100563. https://doi.org/10.1016/j.apmt.2020.100563

[231] Shi, Z., Sun, G., Yuan, R., Chen, W., Wang, Z., Zhang, L., Zhan, K., Zhu, M., Yang, J., Zhao, B., Journal of Materials Science and Technology, 99, 2022, 260-269. https://doi.org/10.1016/j.jmst.2021.05.040

[232] Jiang, Y., Li, X., Liu, F., Wang, B., Zhou, W., Dong, S., Fan, X., Applied Surface Science, 576, 2022, 151801. https://doi.org/10.1016/j.apsusc.2021.151801

[233] Kharangarh, P.R., Ravindra, N.M., Rawal, R., Singh, A., Gupta, V., Journal of Alloys and Compounds, 876, 2021, 159990. https://doi.org/10.1016/j.jallcom.2021.159990

[234] Yan, K., Wu, J., Wang, Y.Y., Liu, N.N., Li, J.T., Gao, Y.P., Hou, Z.Q., Chemical Papers, 74[2] 2020, 591-599. https://doi.org/10.1007/s11696-019-00899-3

[235] Zhang, X., Zhu, M., Ouyang, T., Chen, Y., Yan, J., Zhu, K., Ye, K., Wang, G., Cheng, K., Cao, D., Chemical Engineering Journal, 360, 2019, 171-179. https://doi.org/10.1016/j.cej.2018.11.206

[236] Sahoo, S., Zhang, S., Shim, J.J., Electrochimica Acta, 216, 2016, 386-396. https://doi.org/10.1016/j.electacta.2016.09.030

[237] Hu, N., Huang, L., Gong, W., Shen, P.K., ACS Sustainable Chemistry and Engineering, 6[12] 2018, 16933-16940. https://doi.org/10.1021/acssuschemeng.8b04265

[238] Ghosh, D., Giri, S., Moniruzzaman, M., Basu, T., Mandal, M., Das, C.K., Dalton Transactions, 43[28] 2014, 11067-11076. https://doi.org/10.1039/C4DT00672K

[239] Liu, X., Zhang, K., Yang, B., Song, W., Liu, Q., Jia, F., Qin, S., Chen, W., Zhang, Z., Li, J., Materials Letters, 164, 2016, 401-404. https://doi.org/10.1016/j.matlet.2015.11.051

[240] Muthu, D., Vargheese, S., Haldorai, Y., Rajendra Kumar, R.T., Materials Science in Semiconductor Processing, 135, 2021, 106078. https://doi.org/10.1016/j.mssp.2021.106078

[241] Chen, T., Xiang, C., Zou, Y., Xu, F., Sun, L., Energy and Fuels, 35[12] 2021, 10250-10261. https://doi.org/10.1021/acs.energyfuels.1c00913

[242] Xuan, H., Wang, R., Yang, J., Zhang, G., Liang, X., Li, Y., Xie, Z., Han, P., Journal of Materials Science, 56[15] 2021, 9419-9433. https://doi.org/10.1007/s10853-021-05902-5

[243] Chebrolu, V.T., Balakrishnan, B., Raja, S.A., Kim, H.J., Dalton Transactions, 49[28] 2020, 9762-9772. https://doi.org/10.1039/D0DT00624F

[244] Yuan, J., Yao, D., Jiang, L., Tao, Y., Che, J., He, G., Chen, H., ACS Applied Energy Materials, 3[2] 2020, 1794-1803. https://doi.org/10.1021/acsaem.9b02238

[245] Bu, Y., Wang, S., Jin, H., Zhang, W., Lin, J., Wang, J., Journal of the Electrochemical Society, 159[7] 2012, A990-A994. https://doi.org/10.1149/2.036207jes

[246] Jing, M., Wang, C., Hou, H., Wu, Z., Zhu, Y., Yang, Y., Jia, X., Zhang, Y., Ji, X., Journal of Power Sources, 298, 2015, 241-248. https://doi.org/10.1016/j.jpowsour.2015.08.039

[247] Liu, M., Chang, J., Bai, Y., Sun, J., RSC Advances, 5[111] 2015, 91389-91397. https://doi.org/10.1039/C5RA18976D

[248] Gawali, S.R., Dubal, D.P., Deonikar, V.G., Patil, S.S., Patil, S.D., Gomez-Romero, P., Patil, D.R., Pant, J., ChemistrySelect, 1[13] 2016, 3471-3478. https://doi.org/10.1002/slct.201600566

[249] Li, Q., Wei, Q., Xie, L., Chen, C., Lu, C., Su, F.Y., Zhou, P., RSC Advances, 6[52] 2016, 46548-46557. https://doi.org/10.1039/C6RA04998B

[250] Qiu, Z., He, D., Wang, Y., Wu, H., Li, J., Journal of Alloys and Compounds, 727, 2017, 1189-1202. https://doi.org/10.1016/j.jallcom.2017.08.260

[251] Jiao, X., Hao, Q., Xia, X., Lei, W., Ouyang, Y., Ye, H., Mandler, D., ChemSusChem, 11[5] 2018, 907-915. https://doi.org/10.1002/cssc.201702283

[252] Wang, Y., Pan, D., Zhang, M., Ma, T., Liu, H., Yan, Z., Xie, J., Journal of Alloys and Compounds, 765, 2018, 480-488. https://doi.org/10.1016/j.jallcom.2018.06.222

[253] Zhang, Z., Huo, H., Gao, J., Yu, Z., Ran, F., Guo, L., Lou, S., Mu, T., Yin, X., Wang, Q., Yin, G., Journal of Alloys and Compounds, 801, 2019, 158-165. https://doi.org/10.1016/j.jallcom.2019.06.073

[254] Huang, C., Hao, C., Zheng, W., Zhou, S., Yang, L., Wang, X., Jiang, C., Zhu, L., Applied Surface Science, 505, 2020, 144589. https://doi.org/10.1016/j.apsusc.2019.144589

[255] Gao, X., Zhang, H., Guo, E., Yao, F., Wang, Z., Yue, H., Microchemical Journal, 164, 2021, 105979. https://doi.org/10.1016/j.microc.2021.105979

[256] Saranya, P.E., Selladurai, S., Journal of Materials Science, 53[23] 2018, 16022-16046. https://doi.org/10.1007/s10853-018-2742-1

[257] Low, W.H., Khiew, P.S., Lim, S.S., Siong, C.W., Ezeigwe, E.R., Journal of Alloys and Compounds, 768, 2018, 995-1005. https://doi.org/10.1016/j.jallcom.2018.07.139

[258] Zhang, J., Jiang, J., Li, H., Zhao, X.S., Energy and Environmental Science, 4[10] 2011, 4009-4015. https://doi.org/10.1039/c1ee01354h

[259] Choi, B.G., Chang, S.J., Kang, H.W., Park, C.P., Kim, H.J., Hong, W.H., Lee, S., Huh, Y.S., Nanoscale, 4[16] 2012, 4983-4988. https://doi.org/10.1039/c2nr30991b

[260] Wang, W., Guo, S., Lee, I., Ahmed, K., Zhong, J., Favors, Z., Zaera, F., Ozkan, M., Ozkan, C.S., Scientific Reports, 4, 2014, 4452. https://doi.org/10.1038/srep04452

[261] Wang, R., Jia, P., Yang, Y., An, N., Zhang, Y., Wu, H., Hu, Z., Chinese Journal of Chemistry, 34[1] 2016, 114-122. https://doi.org/10.1002/cjoc.201500595

[262] Meng, Y., Wang, L., Xiao, H., Ma, Y., Chao, L., Xie, Q., RSC Advances, 6[40] 2016, 33666-33675. https://doi.org/10.1039/C6RA03615E

[263] Cho, S., Kim, J., Jo, Y., Ahmed, A.T.A., Chavan, H.S., Woo, H., Inamdar, A.I., Gunjakar, J.L., Pawar, S.M., Park, Y., Kim, H., Im, H., Journal of Alloys and Compounds, 725, 2017, 108-114. https://doi.org/10.1016/j.jallcom.2017.07.135

[264] Thangappan, R., Arivanandhan, M., Kumar, R.D, Jayavel, R., Journal of Physics and Chemistry of Solids, 121, 2018, 339-349. https://doi.org/10.1016/j.jpcs.2018.05.049

[265] Li, M., He, H., Applied Surface Science, 439, 2018, 612-622. https://doi.org/10.1016/j.apsusc.2018.01.064

[266] Male, U., Uppugalla, S., Srinivasan, P., Journal of Solid State Electrochemistry, 19[11] 2015, 3381-3388. https://doi.org/10.1007/s10008-015-2978-5

[267] Ravi, B., Rajender, B., Palaniappan, S., International Journal of Polymeric Materials and Polymeric Biomaterials, 65[16] 2016, 835-840. https://doi.org/10.1080/00914037.2016.1171221

[268] Mazloum-Ardakani, M., Sabaghian, F., Naderi, H., Ebadi, A., Mohammadian-Sarcheshmeh, H., Journal of Solid State Electrochemistry, 24[3] 2020, 571-582. https://doi.org/10.1007/s10008-019-04457-5

[269] Wang, W., Hao, Q., Lei, W., Xia, X., Wang, X., RSC Advances, 2[27] 2012, 10268-10274. https://doi.org/10.1039/c2ra21292g

[270] Wang, W., Lei, W., Yao, T., Xia, X., Huang, W., Hao, Q., Wang, X., Electrochimica Acta, 108, 2013, 118-126. https://doi.org/10.1016/j.electacta.2013.07.012

[271] Velmurugan, V., Srinivasarao, U., Ramachandran, R., Saranya, M., Grace, A.N., Materials Research Bulletin, 84, 2016, 145-151. https://doi.org/10.1016/j.materresbull.2016.07.015

[272] Kim, D.W., Jung, S.M., Jung, H.Y., Journal of Power Sources, 448, 2020, 227422. https://doi.org/10.1016/j.jpowsour.2019.227422

[273] Vandana, M., Nagaraju, Y.S., Ganesh, H., Veeresh, S., Vijeth, H., Basappa, M., Devendrappa, H., RSC Advances, 11[45] 2021, 27801-27811. https://doi.org/10.1039/D1RA03423E

[274] Fu, M., Zhang, Z., Zhu, Z., Zhuang, Q., Chen, W., Yu, H., Liu, Q., Journal of Colloid and Interface Science, 588, 2021, 795-803. https://doi.org/10.1016/j.jcis.2020.11.114

[275] Xiang, C., Li, M., Zhi, M., Manivannan, A., Wu, N., Journal of Materials Chemistry, 22[36] 2012, 19161-19167. https://doi.org/10.1039/c2jm33177b

[276] Kim, H., Cho, M.Y., Kim, M.H., Park, K.Y., Gwon, H., Lee, Y., Roh, K.C., Kang, K., Advanced Energy Materials, 3[11] 2013, 1500-1506.

https://doi.org/10.1002/aenm.201300467

[277] Ates, M., Bayrak, Y., Yoruk, O., Caliskan, S., Journal of Alloys and Compounds, 728, 2017, 541-551. https://doi.org/10.1016/j.jallcom.2017.08.298

[278] Du, J., Zheng, C., Lv, W., Deng, Y., Pan, Z., Kang, F., Yang, Q.H., Advanced Materials Interfaces, 4[11] 2017, 1700004. https://doi.org/10.1002/admi.201700004

[279] Yue, H.Y., Guan, E.H., Gao, X., Yao, F., Wang, W.Q., Zhang, T., Wang, Z., Song, S.S., Zhang, H.J., ChemistrySelect, 3[44] 2018, 12455-12460. https://doi.org/10.1002/slct.201803149

[280] Yang, S., Li, Y., Sun, J., Cao, B., Journal of Power Sources, 431, 2019, 220-225. https://doi.org/10.1016/j.jpowsour.2019.05.016

[281] Sahoo, R., Lee, T.H., Pham, D.T., Luu, T.H.T., Lee, Y.H., ACS Nano, 13[9] 2019, 10776-10786. https://doi.org/10.1021/acsnano.9b05605

[282] Perera, S.D., Liyanage, A.D., Nijem, N., Ferraris, J.P., Chabal, Y.J., Balkus, K.J., Journal of Power Sources, 230, 2013, 130-137. https://doi.org/10.1016/j.jpowsour.2012.11.118

[283] Shakir, I., Ali, Z., Bae, J., Park, J., Kang, D.J., Nanoscale, 6[8] 2014, 4125-4130. https://doi.org/10.1039/c3nr06820j

[284] Wu, Y., Gao, G., Wu, G., Journal of Materials Chemistry A, 3[5] 2015, 1828-1832. https://doi.org/10.1039/C4TA05537C

[285] Choudhury, A., Bonso, J.S., Wunch, M., Yang, K.S., Ferraris, J.P., Yang, D.J., Journal of Power Sources, 287, 2015, 283-290. https://doi.org/10.1016/j.jpowsour.2015.04.062

[286] Deng, L., Gao, Y., Ma, Z., Fan, G., Journal of Colloid and Interface Science, 505, 2017, 556-565. https://doi.org/10.1016/j.jcis.2017.06.048

[287] Asen, P., Shahrokhian, S., Iraji zad, A., International Journal of Hydrogen Energy, 42[33] 2017, 21073-21085. https://doi.org/10.1016/j.ijhydene.2017.07.008

[288] Bi, W., Gao, G., Wu, Y., Yang, H., Wang, J., Zhang, Y., Liang, X., Liu, Y., Wu, G., RSC Advances, 7[12] 2017, 7179-7187. https://doi.org/10.1039/C6RA25077G

[289] Ndiaye, N.M., Ngom, B.D., Sylla, N.F., Masikhwa, T.M., Madito, M.J., Momodu, D., Ntsoane, T., Manyala, N., Journal of Colloid and Interface Science, 532, 2018, 395-406. https://doi.org/10.1016/j.jcis.2018.08.010

[290] Ahirrao, D.J., Mohanapriya, K., Jha, N., Materials Research Bulletin, 108, 2018,

73-82. https://doi.org/10.1016/j.materresbull.2018.08.028

[291] Santhosh, R., Raman, S.R.S., Krishna, S.M., Ravuri, S.S., Sandhya, V., Ghosh, S., Sahu, N.K., Punniyakoti, S., Karthik, M., Kollu, P., Jeong, S.K., Grace, A.N., Electrochimica Acta, 276, 2018, 284-292. https://doi.org/10.1016/j.electacta.2018.04.142

[292] Boruah, B.D., Nandi, S., Misra, A., ACS Applied Energy Materials, 1[4] 2018, 1567-1574. https://doi.org/10.1021/acsaem.7b00358

[293] Viswanathan, A., Prakashaiah, B.G., Subburaj, V., Shetty, A.N., Journal of Colloid and Interface Science, 545, 2019, 82-93. https://doi.org/10.1016/j.jcis.2019.03.013

[294] Hu, T., Liu, Y., Zhang, Y., Chen, M., Zheng, J., Tang, J., Meng, C., Journal of Colloid and Interface Science, 531, 2018, 382-393. https://doi.org/10.1016/j.jcis.2018.07.060

[295] Nayak, A.K., Das, A.K., Pradhan, D., ACS Sustainable Chemistry and Engineering, 5[11] 2017, 10128-10138. https://doi.org/10.1021/acssuschemeng.7b02135

[296] Patil, A.M., Wang, J., Li, S., Hao, X., Du, X., Wang, Z., Hao, X., Abudula, A., Guan, G., Chemical Engineering Journal, 421, 2021, 127883. https://doi.org/10.1016/j.cej.2020.127883

[297] Ashraf, M., Shah, S.S., Khan, I., Aziz, M.A., Ullah, N., Khan, M., Adil, S.F., Liaqat, Z., Usman, M., Tremel, W., Tahir, M.N., Chemistry - A European Journal, 27[23] 2021, 6973-6984. https://doi.org/10.1002/chem.202005156

[298] Lee, I., Jeong, G.H., Lim, J., Kim, S.W., Yoon, S., Journal of Alloys and Compounds, 758, 2018, 99-107. https://doi.org/10.1016/j.jallcom.2018.05.059

[299] Sahoo, S., Shim, J.J., ACS Sustainable Chemistry and Engineering, 5[1] 2017, 241-251. https://doi.org/10.1021/acssuschemeng.6b01367

[300] Sahoo, S., Shim, J.J., Journal of Industrial and Engineering Chemistry, 54, 2017, 205-217. https://doi.org/10.1016/j.jiec.2017.05.035

[301] Ahuja, P., Sharma, R.K., Singh, G., Journal of Materials Chemistry A, 3[9] 2015, 4931-4937. https://doi.org/10.1039/C4TA05865H

[302] Reddy, B.J., Vickraman, P., Justin, A.S., Physica Status Solidi A, 216[2] 2019, 1800595. https://doi.org/10.1002/pssa.201800595

[303] Li, J., Sun, Y., Li, D., Yang, H., Zhang, X., Lin, B., Journal of Alloys and Compounds, 708, 2017, 787-795. https://doi.org/10.1016/j.jallcom.2017.03.062

[304] Zhang, M., Ma, T., Wang, Y., Pan, D., Xie, J., Journal of Materials Science: Materials in Electronics, 29[8] 2018, 6991-7001. https://doi.org/10.1007/s10854-018-8686-z

[305] Samuel, E., Londhe, P.U., Joshi, B., Kim, M.W., Kim, K., Swihart, M.T., Chaure, N.B., Yoon, S.S., Journal of Alloys and Compounds, 741, 2018, 781-791. https://doi.org/10.1016/j.jallcom.2017.12.320

[306] Halder, L., Kumar Das, A., Maitra, A., Bera, A., Paria, S., Karan, S.K., Si, S.K., Ojha, S., De, A., Khatua, B.B., New Journal of Chemistry, 44[3] 2020, 1063-1075. https://doi.org/10.1039/C9NJ05546K

[307] Low, W.H., Khiew, P.S., Lim, S.S., Siong, C.W., Chia, C.H., Ezeigwe, E.R., Journal of Alloys and Compounds, 784, 2019, 847-858. https://doi.org/10.1016/j.jallcom.2019.01.137

[308] Iqbal, M.F., Mahmood-Ul-Hassan, Ashiq, M.N., Iqbal, S., Bibi, N., Parveen, B., Electrochimica Acta, 246, 2017, 1097-1103. https://doi.org/10.1016/j.electacta.2017.06.123

[309] Yang, X., Sun, H., Zan, P., Zhao, L., Lian, J., Journal of Materials Chemistry A, 4[48] 2016, 18857-18867. https://doi.org/10.1039/C6TA07898B

[310] Ma, T., Zhang, M., Liu, H., Wang, Y., Electrochimica Acta, 322, 2019, 134762. https://doi.org/10.1016/j.electacta.2019.134762

[311] Lin, J., Yan, S., Liu, P., Chang, X., Yao, L., Lin, H., Lu, D., Han, S., Research on Chemical Intermediates, 44[7] 2018, 4503-4518. https://doi.org/10.1007/s11164-018-3400-6

[312] Gao, Z., Chen, C., Chang, J., Chen, L., Wang, P., Wu, D., Xu, F., Guo, Y., Jiang, K., Electrochimica Acta, 281, 2018, 394-404. https://doi.org/10.1016/j.electacta.2018.05.194

[313] Chang, J., Zang, S., Liang, W., Wu, D., Lian, Z., Xu, F., Jiang, K., Gao, Z., Journal of Colloid and Interface Science, 590, 2021, 114-124. https://doi.org/10.1016/j.jcis.2021.01.035

[314] De, B., Kuila, T., Kim, N.H., Lee, J.H., Carbon, 122, 2017, 247-257. https://doi.org/10.1016/j.carbon.2017.06.076

[315] Zhao, T., Yang, W., Zhao, X., Peng, X., Hu, J., Tang, C., Li, T., Composites B, 150, 2018, 60-67. https://doi.org/10.1016/j.compositesb.2018.05.058

[316] Jiang, D., Liang, H., Yang, W., Liu, Y., Cao, X., Zhang, J., Li, C., Liu, J., Gooding,

J.J., Carbon, 146, 2019, 557-567. https://doi.org/10.1016/j.carbon.2019.02.045

[317] Li, H., Xuan, H., Gao, J., Liang, T., Han, X., Guan, Y., Yang, J., Han, P., Du, Y., Electrochimica Acta, 312, 2019, 213-223. https://doi.org/10.1016/j.electacta.2019.05.008

[318] Ji, Z., Li, N., Xie, M., Shen, X., Dai, W., Liu, K., Xu, K., Zhu, G., Electrochimica Acta, 334, 2020, 135632. https://doi.org/10.1016/j.electacta.2020.135632

[319] Wu, C.L., Chen, D.H., Journal of Alloys and Compounds, 872, 2021, 159702. https://doi.org/10.1016/j.jallcom.2021.159702

[320] Zhang, Q., Xu, C., Lu, B., Electrochimica Acta, 132, 2014, 180-185. https://doi.org/10.1016/j.electacta.2014.03.111

[321] Ghosh, D., Das, C.K., ACS Applied Materials and Interfaces, 7[2] 2015, 1122-1131. https://doi.org/10.1021/am506738y

[322] Patil, S.J., Kim, J.H., Lee, D.W., Journal of Power Sources, 342, 2017, 652-665. https://doi.org/10.1016/j.jpowsour.2016.12.096

[323] Lin, T.W., Dai, C.S., Tasi, T.T., Chou, S.W., Lin, J.Y., Shen, H.H., Chemical Engineering Journal, 279, 2015, 241-249. https://doi.org/10.1016/j.cej.2015.05.011

[324] Wang, X., Su, D., Xiao, Y., Xu, S., Fang, S., Cao, S., Electrochimica Acta, 293, 2019, 419-425. https://doi.org/10.1016/j.electacta.2018.10.059

[325] Sarkar, S., Howli, P., Das, B., Das, N.S., Samanta, M., Das, G.C., Chattopadhyay, K.K., ACS Applied Materials and Interfaces, 9[27] 2017, 22652-22664. https://doi.org/10.1021/acsami.7b00437

[326] Tang, Y., Chen, T., Yu, S., Qiao, Y., Mu, S., Hu, J., Gao, F., Journal of Materials Chemistry A, 3[24] 2015, 12913-12919. https://doi.org/10.1039/C5TA02480C

[327] Zhang, C., Lin, Z., Huang, C., Zheng, B., Li, Y., Wang, J., Deng, M., Tang, S., Du, Y., ACS Applied Energy Materials, 2[9] 2019, 6599-6607. https://doi.org/10.1021/acsaem.9b01149

[328] Huang, K.J., Wang, L., Liu, Y.J., Liu, Y.M., Wang, H.B., Gan, T., Wang, L.L., International Journal of Hydrogen Energy, 38[32] 2013, 14027-14034. https://doi.org/10.1016/j.ijhydene.2013.08.112

[329] Huang, K.J., Wang, L., Liu, Y.J., Wang, H.B., Liu, Y.M., Wang, L.L., Electrochimica Acta, 109, 2013, 587-594. https://doi.org/10.1016/j.electacta.2013.07.168

[330] Da Silveira Firmiano, E.G., Rabelo, A.C., Dalmaschio, C.J., Pinheiro, A.N., Pereira, E.C., Schreiner, W.H., Leite, E.R., Advanced Energy Materials, 4[6] 2014, 1301380. https://doi.org/10.1002/aenm.201301380

[331] Thangappan, R., Kalaiselvam, S., Elayaperumal, A., Jayavel, R., Arivanandhan, M., Karthikeyan, R., Hayakawa, Y., Dalton Transactions, 45[6] 2016, 2637-2646. https://doi.org/10.1039/C5DT04832J

[332] Zhou, R., Han, C.J., Wang, X.M., Journal of Power Sources, 352, 2017, 99-110. https://doi.org/10.1016/j.jpowsour.2017.03.134

[333] Palsaniya, S., Nemade, H.B., Dasmahapatra, A.K., Polymer, 150, 2018, 150-158. https://doi.org/10.1016/j.polymer.2018.07.018

[334] Xu, G., Chen, S., Liu, Y., Fan, W., Functional Materials Letters, 11[4] 2018, 1850074. https://doi.org/10.1142/S1793604718500741

[335] Lin, T.W., Sadhasivam, T., Wang, A.Y., Chen, T.Y., Lin, J.Y., Shao, L.D., ChemElectroChem, 5[7] 2018, 1024-1031. https://doi.org/10.1002/celc.201800043

[336] Han, C., Tian, Z., Dou, H., Wang, X., Yang, X., Chinese Chemical Letters, 29[4] 2018, 606-611. https://doi.org/10.1016/j.cclet.2018.01.017

[337] Li, H., Jiang, N., Deng, Q., Wang, X., ChemistrySelect, 4[43] 2019, 12815-12823. https://doi.org/10.1002/slct.201903517

[338] Khoh, W.H., Wee, B.H., Hong, J.D., Colloids and Surfaces A, 581, 2019, 123815. https://doi.org/10.1016/j.colsurfa.2019.123815

[339] Zhou, C., Hong, M., Yang, Y., Yang, C., Hu, N., Zhang, L., Yang, Z., Zhang, Y., Journal of Power Sources, 438, 2019, 227044. https://doi.org/10.1016/j.jpowsour.2019.227044

[340] Sarmah, D., Kumar, A., Electrochimica Acta, 312, 2019, 392-410. https://doi.org/10.1016/j.electacta.2019.04.174

[341] Xue, T., Yang, Y., Yan, X.H., Zou, Z.L., Han, F., Yang, Z., Materials Research Express, 6[9] 2019, 095029. https://doi.org/10.1088/2053-1591/ab2ebc

[342] Wang, C., Zhai, S., Yuan, Z., Chen, J., Zhang, X., Huang, Q., Wang, Y., Liao, X., Wei, L., Chen, Y., Electrochimica Acta, 305, 2019, 493-501. https://doi.org/10.1016/j.electacta.2019.03.084

[343] Vikraman, D., Karuppasamy, K., Hussain, S., Kathalingam, A., Sanmugam, A., Jung, J., Kim, H.S., Composites B, 161, 2019, 555-563. https://doi.org/10.1016/j.compositesb.2018.12.143

[344] Zhao, J., Gao, L., Wang, Z., Wang, S., Xu, R., Journal of Alloys and Compounds, 887, 2021, 161514. https://doi.org/10.1016/j.jallcom.2021.161514

[345] Li, Z., Wang, X., Xu, M., Yin, Z., Zhao, J., Journal of Alloys and Compounds, 894, 2022, 162492. https://doi.org/10.1016/j.jallcom.2021.162492

[346] Bokhari, S.W., Wei, S., Gao, W., Electrochimica Acta, 398, 2021, 139300. https://doi.org/10.1016/j.electacta.2021.139300

[347] Kumar, S., Riyajuddin, S., Afshan, M., Aziz, S.K., Maruyama, T., Ghosh, K., Journal of Physical Chemistry Letters, 12[28] 2021, 6574-6581. https://doi.org/10.1021/acs.jpclett.1c01553

[348] Hao, J., Liu, H., Han, S., Lian, J., ACS Applied Nano Materials, 4[2] 2021, 1330-1339. https://doi.org/10.1021/acsanm.0c02899

[349] Fu, M., Zhu, Z., Chen, W., Yu, H., Liu, Q., Journal of Materials Science, 55[34] 2020, 16385-16393. https://doi.org/10.1007/s10853-020-05201-5

[350] Zhao, S., Xu, W., Yang, Z., Zhang, X., Zhang, Q., Electrochimica Acta, 331, 2020, 135265. https://doi.org/10.1016/j.electacta.2019.135265

[351] Chen, Y., Bai, J., Yang, D., Sun, P., Li, X., Electrochimica Acta, 330, 2020, 135205. https://doi.org/10.1016/j.electacta.2019.135205

[352] Zhang, M., Song, Z., Liu, H., Wang, A., Shao, S., Journal of Colloid and Interface Science, 584, 2021, 418-428. https://doi.org/10.1016/j.jcis.2020.10.005

[353] Wang, F., Li, G., Zhou, Q., Zheng, J., Yang, C., Wang, Q., Applied Surface Science, 425, 2017, 180-187. https://doi.org/10.1016/j.apsusc.2017.07.016

[354] Beka, L.G., Li, X., Liu, W., Scientific Reports, 7[1] 2017, 2105. https://doi.org/10.1038/s41598-017-02309-8

[355] Li, B., Tian, Z., Li, H., Yang, Z., Wang, Y., Wang, X., Electrochimica Acta, 314, 2019, 32-39. https://doi.org/10.1016/j.electacta.2019.05.040

[356] Zhai, R., Xiao, Y., Ding, T., Wu, Y., Chen, S., Wei, W., Journal of Alloys and Compounds, 845, 2020, 156164. https://doi.org/10.1016/j.jallcom.2020.156164

[357] Guo, R., Dang, L., Liu, Z., Lei, Z., Colloids and Surfaces A, 602, 2020, 125110. https://doi.org/10.1016/j.colsurfa.2020.125110

[358] Van Hoa, N., Dat, P.A., Van Hieu, N., Le, T.N., Minh, N.C., Van Tang, N., Nga, D.T., Ngoc, T.Q., Diamond and Related Materials, 106, 2020, 107850. https://doi.org/10.1016/j.diamond.2020.107850

[359] Hsiang, H.I., She, C.H., Chung, S.H., Ceramics International, 47[18] 2021, 25942-25950. https://doi.org/10.1016/j.ceramint.2021.05.325

[360] Poompiew, N., Pattananuwat, P., Potiyaraj, P., RSC Advances, 11[40] 2021, 25057-25067. https://doi.org/10.1039/D1RA03607F

[361] Yu, J., Li, H., Shi, X., Sun, Z., Journal of Electronic Materials, 50[6] 2021, 3095-3104. https://doi.org/10.1007/s11664-021-08837-4

[362] Xu, X., Liang, L., Zhang, Z., Xiong, R., Zhang, X., Zhao, Y., Qiao, S., Zhang, Y., Diamond and Related Materials, 112, 2021, 108240. https://doi.org/10.1016/j.diamond.2021.108240

[363] Jothi, P.R., Salunkhe, R.R., Pramanik, M., Kannan, S., Yamauchi, Y., RSC Advances, 6[25] 2016, 21246-21253. https://doi.org/10.1039/C5RA26946F

[364] Reddy, B.J., Vickraman, P., Justin, A.S., Journal of Materials Science, 54[8] 2019, 6361-6373. https://doi.org/10.1007/s10853-018-03314-6

[365] Lin, T.W., Dai, C.S., Hung, K.C., Scientific Reports, 4, 2014, 7274.

[366] Qi, J., Chang, Y., Sui, Y., He, Y., Meng, Q., Wei, F., Zhao, Y., Jin, Y., Particle and Particle Systems Characterization, 34[12] 2017, 1700196. https://doi.org/10.1002/ppsc.201700196

[367] Zhang, C., Huang, Y., Tang, S., Deng, M., Du, Y., ACS Energy Letters, 2[4] 2017, 759-768. https://doi.org/10.1021/acsenergylett.7b00078

[368] Qi, M., Zhu, W., Lu, Z., Zhang, H., Ling, Y., Ou, X., Journal of Materials Science: Materials in Electronics, 31[15] 2020, 12536-12545. https://doi.org/10.1007/s10854-020-03804-x

[369] Hu, Q., Zou, X., Huang, Y., Wei, Y., YaWang, Chen, F., Xiang, B., Wu, Q., Li, W., Journal of Colloid and Interface Science, 559, 2020, 115-123. https://doi.org/10.1016/j.jcis.2019.10.010

[370] Wang, R., Xuan, H., Zhang, G., Li, H., Guan, Y., Liang, X., Zhang, S., Wu, Z., Han, P., Wu, Y., Applied Surface Science, 526, 2020, 146641. https://doi.org/10.1016/j.apsusc.2020.146641

[371] Nandhini, S., Muralidharan, G., Electrochimica Acta, 365, 2021, 137367. https://doi.org/10.1016/j.electacta.2020.137367

[372] Lonkar, S.P., Pillai, V.V., Patole, S.P., Alhassan, S.M., ACS Applied Energy Materials, 3[5] 2020, 4995-5005. https://doi.org/10.1021/acsaem.0c00519

[373] Iqbal, M.F., Ashiq, M.N., Razaq, A., Saleem, M., Parveen, B., Hassan, M.U.,

Electrochimica Acta, 273, 2018, 136-144.
https://doi.org/10.1016/j.electacta.2018.04.014

[374] Ubale, S.B., Kale, S.B., Mane, V.J., Patil, U.M., Lokhande, C.D., Journal of Electroanalytical Chemistry, 897, 2021, 115589.
https://doi.org/10.1016/j.jelechem.2021.115589

[375] Tong, H., Bai, W., Yue, S., Gao, Z., Lu, L., Shen, L., Dong, S., Zhu, J., He, J., Zhang, X., Journal of Materials Chemistry A, 4[29] 2016, 11256-11263.
https://doi.org/10.1039/C6TA02249A

[376] Li, X., Cao, J., Yang, L., Wei, M., Liu, X., Liu, Q., Hong, Y., Zhou, Y., Yang, J., Dalton Transactions, 48[7] 2019, 2442-2454.
https://doi.org/10.1039/C8DT04097D

[377] Iqbal, M.F., Mahmood-Ul-Hassan, Razaq, A., Ashiq, M.N., Kaneti, Y.V., Azhar, A.A., Yasmeen, F., Joya, K.S., Abbass, S., ChemElectroChem, 5[18] 2018, 2636-2644. https://doi.org/10.1002/celc.201800633

[378] Bahaa, A., Abdelkareem, M.A., Al naqbi, H., Yousef Mohamed, A., Shinde, P.A., Yousef, B.A.A., Sayed, E.T., Alawadhi, H., Chae, K.J., Al-Asheh, S., Olabi, A.G., Journal of Colloid and Interface Science, 608, 2022, 711-719.
https://doi.org/10.1016/j.jcis.2021.09.136

[379] Ghosh, S., Samanta, P., Murmu, N.C., Kuila, T., Journal of Alloys and Compounds, 835, 2020, 155432. https://doi.org/10.1016/j.jallcom.2020.155432

[380] Meng, A., Shen, T., Huang, T., Song, G., Li, Z., Tan, S., Zhao, J., Science China Materials, 63[2] 2020, 229-239. https://doi.org/10.1007/s40843-019-9587-5

[381] Marri, S.R., Ratha, S., Rout, C.S., Behera, J.N., Chemical Communications, 53[1] 2017, 228-231. https://doi.org/10.1039/C6CC08035A

[382] Gopi, C.V.V.M., Reddy, A.E., Bak, J.S., Cho, I.H., Kim, H.J., Materials Letters, 223, 2018, 57-60. https://doi.org/10.1016/j.matlet.2018.04.023

[383] Jiao, L., Pan, X., Xi, Y., Li, J., Cao, J., Guo, Q., Han, W., Journal of Materials Science - Materials in Electronics, 30[1] 2019, 159-166.
https://doi.org/10.1007/s10854-018-0277-5

[384] Xie, H., Tang, S., Zhu, J., Vongehr, S., Meng, X., Journal of Materials Chemistry A, 3[36] 2015, 18505-18513. https://doi.org/10.1039/C5TA05129K

[385] Wang, D., Zhang, Y., Yang, L., Fan, G., Lin, Y., Li, F., Journal of Materials Science: Materials in Electronics, 31[8] 2020, 6467-6478.

https://doi.org/10.1007/s10854-020-03202-3

[386] Zhao, B., Chen, D., Xiong, X., Song, B., Hu, R., Zhang, Q., Rainwater, B.H., Waller, G.H., Zhen, D., Ding, Y., Chen, Y., Qu, C., Dang, D., Wong, C.P., Liu, M., Energy Storage Materials, 7, 2017, 32-39. https://doi.org/10.1016/j.ensm.2016.11.010

[387] Xie, L.J., Wu, J.F., Chen, C.M., Zhang, C.M., Wan, L., Wang, J.L., Kong, Q.Q., Lv, C.X., Li, K.X., Sun, G.H., Journal of Power Sources, 242, 2013, 148-156. https://doi.org/10.1016/j.jpowsour.2013.05.081

[388] Cheng, Q., Tang, J., Shinya, N., Qin, L.C., Science and Technology of Advanced Materials, 15[1] 2014, 014206. https://doi.org/10.1088/1468-6996/15/1/014206

[389] Li, N., Zhi, C., Zhang, H., Electrochimica Acta, 220, 2016, 618-627. https://doi.org/10.1016/j.electacta.2016.10.068

[390] Zhao, C., Ren, F., Xue, X., Zheng, W., Wang, X., Chang, L., Journal of Electroanalytical Chemistry, 782, 2016, 98-102. https://doi.org/10.1016/j.jelechem.2016.10.023

[391] Rezaei, B., Jahromi, A.R.T., Ensafi, A.A., International Journal of Hydrogen Energy, 42[26] 2017, 16538-16546. https://doi.org/10.1016/j.ijhydene.2017.05.193

[392] Jin, E.M., Lee, H.J., Jun, H.B., Jeong, S.M., Korean Journal of Chemical Engineering, 34[3] 2017, 885-891. https://doi.org/10.1007/s11814-016-0323-z

[393] Cheng, J.P., Liu, L., Ma, K.Y., Wang, X., Li, Q.Q., Wu, J.S., Liu, F., Journal of Colloid and Interface Science, 486, 2017, 344-350. https://doi.org/10.1016/j.jcis.2016.09.064

[394] Rahimi, S.A., Norouzi, P., Ganjali, M.R., RSC Advances, 8[47] 2018, 26818-26827. https://doi.org/10.1039/C8RA04105A

[395] Rong, Y., Chen, Y., Zheng, J., Zhao, Y., Li, Q., Journal of Colloid and Interface Science, 598, 2021, 1-13. https://doi.org/10.1016/j.jcis.2021.04.029

[396] Pramanik, A., Maiti, S., Mahanty, S., Dalton Transactions, 44[33] 2015, 14604-14612. https://doi.org/10.1039/C5DT01643F

[397] Sun, S., Wang, S., Xia, T., Li, X., Jin, Q., Wu, Q., Wang, L., Wei, Z., Wang, P., Journal of Materials Chemistry A, 3[42] 2015, 20944-20951. https://doi.org/10.1039/C5TA04851F

[398] Arunachalam, S., Kirubasankar, B., Murugadoss, V., Vellasamy, D., Angaiah, S.,

New Journal of Chemistry, 42[4] 2018, 2923-2932.
https://doi.org/10.1039/C7NJ04335J

[399] Wang, D., Wei, A., Tian, L., Mensah, A., Li, D., Xu, Y., Wei, Q., Applied Surface Science, 483, 2019, 593-600. https://doi.org/10.1016/j.apsusc.2019.03.345

[400] Wang, Y., Zhou, D., Zhao, D., Hou, M., Wang, C., Xia, Y., Journal of the Electrochemical Society, 160[1] 2013, A98-A104.
https://doi.org/10.1149/2.012302jes

[401] Xu, Y., Huang, X., Lin, Z., Zhong, X., Huang, Y., Duan, X., Nano Research, 6[1] 2013, 65-76. https://doi.org/10.1007/s12274-012-0284-4

[402] Wu, Z., Huang, X.L., Wang, Z.L., Xu, J.J., Wang, H.G., Zhang, X.B., Scientific Reports, 4, 2014, 3669. https://doi.org/10.1038/srep03669

[403] Bag, S., Raj, C.R., Journal of Materials Chemistry A, 2[42] 2014, 17848-17856.
https://doi.org/10.1039/C4TA02937B

[404] Liu, Y., Wang, R., Yan, X., Scientific Reports, 5, 2015, 11095.
https://doi.org/10.1038/srep11271

[405] Yuan, S., Lu, C., Li, Y., Wang, X., ChemElectroChem, 4[11] 2017, 2826-2834.
https://doi.org/10.1002/celc.201700482

[406] Gao, H., Hao, C., Qi, Y., Li, J., Wang, X., Zhou, S., Huang, C., Journal of Alloys and Compounds, 767, 2018, 1048-1056.
https://doi.org/10.1016/j.jallcom.2018.07.181

[407] Xu, H., Wu, J., Liu, J., Chen, Y., Fan, X., Journal of Materials Science: Materials in Electronics, 29[20] 2018, 17234-17244. https://doi.org/10.1007/s10854-018-9817-2

[408] Fan, W., Shi, Y., Gao, W., Sun, Z., Liu, T., ACS Applied Nano Materials, 1[9] 2018, 4435-4441. https://doi.org/10.1021/acsanm.8b00605

[409] Li, J., Hao, C., Zhou, S., Huang, C., Wang, X., Electrochimica Acta, 283, 2018, 467-477. https://doi.org/10.1016/j.electacta.2018.06.155

[410] Cho, E.C., Chang-Jian, C.W., Lee, K.C., Huang, J.H., Ho, B.C., Liu, R.Z., Hsiao, Y.S., Chemical Engineering Journal, 334, 2018, 2058-2067.
https://doi.org/10.1016/j.cej.2017.11.175

[411] Huang, Y., Buffa, A., Deng, H., Sarkar, S., Ouyang, Y., Jiao, X., Hao, Q., Mandler, D., Journal of Power Sources, 439, 2019, 227046.
https://doi.org/10.1016/j.jpowsour.2019.227046

[412] Wen, S., Qin, K., Liu, P., Zhao, N., Shi, C., Ma, L., Liu, E., Journal of Alloys and Compounds, 783, 2019, 625-632. https://doi.org/10.1016/j.jallcom.2018.12.347

[413] Qi, Y., Liu, Y., Zhu, R., Wang, Q., Luo, Y., Zhu, C., Lyu, Y., New Journal of Chemistry, 43[7] 2019, 3091-3098. https://doi.org/10.1039/C8NJ04959A

[414] Zhu, Y., An, S., Sun, X., Cui, J., Zhang, Y., He, W., Vacuum, 172, 2020, 109055. https://doi.org/10.1016/j.vacuum.2019.109055

[415] Hong, Y., Xu, J., Chung, J.S., Choi, W.M., Journal of Materials Science and Technology, 58, 2020, 73-79. https://doi.org/10.1016/j.jmst.2020.03.063

[416] Lo, H.J., Huang, M.C., Lai, Y.H., Chen, H.Y., Materials Chemistry and Physics, 262, 2021, 124306. https://doi.org/10.1016/j.matchemphys.2021.124306

[417] Tong, H., Liu, J., Li, T., Gong, D., Xiao, J., Lu, L., Shen, L., Zhang, X., Nano, 16[2] 2021, 2150013. https://doi.org/10.1142/S1793292021500132

[418] Liu, H., Zhang, M., Ma, T., Wang, Y., Song, Z., Wang, A., Huang, Z., Chemical Engineering Science, 238, 2021, 116613. https://doi.org/10.1016/j.ces.2021.116613

[419] Ma, X., Feng, H., Yan, T., Zhang, L., Liu, X., Cao, S., Dalton Transactions, 50[38] 2021, 13276-13285. https://doi.org/10.1039/D1DT01744F

[420] Yang, C., Dong, L., Chen, Z., Lu, H., Journal of Physical Chemistry C, 118[33] 2014, 18884-18891. https://doi.org/10.1021/jp504741u

[421] Talebi, M., Asen, P., Shahrokhian, S., Ahadian, M.M., Electrochimica Acta, 296, 2019, 130-141. https://doi.org/10.1016/j.electacta.2018.10.192

[422] Mirghni, A.A., Oyedotun, K.O., Mahmoud, B.A., Bello, A., Ray, S.C., Manyala, N., Composites B, 174, 2019, 106953. https://doi.org/10.1016/j.compositesb.2019.106953

[423] Mirghni, A.A., Madito, M.J., Oyedotun, K.O., Masikhwa, T.M., Ndiaye, N.M., Ray, S.J., Manyala, N., RSC Advances, 8[21] 2018, 11608-11621. https://doi.org/10.1039/C7RA12028A

[424] Lee, K.H., Lee, Y.W., Lee, S.W., Ha, J.S., Lee, S.S., Son, J.G., Scientific Reports, 5, 2015, 13696. https://doi.org/10.1038/srep16220

[425] Chen, N., Zhou, J., Kang, Q., Ji, H., Zhu, G., Zhang, Y., Chen, S., Chen, J., Feng, X., Hou, W., Journal of Power Sources, 344, 2017, 185-194. https://doi.org/10.1016/j.jpowsour.2017.01.119

[426] Saha, S., Jana, M., Khanra, P., Samanta, P., Koo, H., Murmu, N.C., Kuila, T., ACS

Applied Materials and Interfaces, 7[26] 2015, 14211-14222.
https://doi.org/10.1021/acsami.5b03562

[427] Ding, Y., Tang, Y., Yang, L., Zeng, Y., Yuan, J., Liu, T., Zhang, S., Liu, C., Luo, S., Journal of Materials Chemistry A, 4[37] 2016, 14307-14315.
https://doi.org/10.1039/C6TA05267C

[428] Yu, Y., Gao, W., Shen, Z., Zheng, Q., Wu, H., Wang, X., Song, W., Ding, K., Journal of Materials Chemistry A, 3[32] 2015, 16633-16641.
https://doi.org/10.1039/C5TA03830H

[429] Balamurugan, J., Karthikeyan, G., Thanh, T.D., Kim, N.H., Lee, J.H., Journal of Power Sources, 308, 2016, 149-157.
https://doi.org/10.1016/j.jpowsour.2016.01.071

[430] Deng, L., Ma, Z., Liu, Z.H., Fan, G., Journal of Alloys and Compounds, 812, 2020, 152087. https://doi.org/10.1016/j.jallcom.2019.152087

[431] Asen, P., Shahrokhian, S., Journal of Physical Chemistry C, 121[12] 2017, 6508-6519. https://doi.org/10.1021/acs.jpcc.7b00534

[432] Kumar, S., Aziz, S.K.T., Kumar, S., Riyajuddin, S., Yaniv, G., Meshi, L., Nessim, G.D., Ghosh, K., Frontiers in Materials, 7, 2020, 30.
https://doi.org/10.3389/fmats.2020.00030

[433] Tan, Q., Chen, X., Wan, H., Zhang, B., Liu, X., Li, L., Wang, C., Gan, Y., Liang, P., Wang, Y., Zhang, J., Wang, H., Miao, L., Jiang, J., van Aken, P.A., Wang, H., Journal of Power Sources, 448, 2020, 227403.
https://doi.org/10.1016/j.jpowsour.2019.227403

[434] Ghosh, D., Giri, S., Dhibar, S., Das, C.K., Electrochimica Acta, 147, 2014, 557-564. https://doi.org/10.1016/j.electacta.2014.09.130

[435] Amutha, B., Sathish, M., Journal of Solid State Electrochemistry, 19[8] 2015, 2311-2320. https://doi.org/10.1007/s10008-015-2867-y

[436] Jana, M., Kumar, J.S., Khanra, P., Samanta, P., Koo, H., Murmu, N.C., Kuila, T., Journal of Power Sources, 303, 2016, 222-233.
https://doi.org/10.1016/j.jpowsour.2015.10.107

[437] Cao, X., Wang, X., Cui, L., Jiang, D., Zheng, Y., Liu, J., Chemical Engineering Journal, 327, 2017, 1085-1092. https://doi.org/10.1016/j.cej.2017.07.010

[438] Ning, J., Xia, M., Wang, D., Feng, X., Zhou, H., Zhang, J., Hao, Y., Nano-Micro Letters, 13[1] 2021, 2. https://doi.org/10.1007/s40820-020-00527-w

[439] Liu, W.W., Feng, Y.Q., Yan, X.B., Chen, J.T., Xue, Q.J., Advanced Functional Materials, 23[33] 2013, 4111-4122. https://doi.org/10.1002/adfm.201203771

[440] Wu, Z.S., Parvez, K., Feng, X., Müllen, K., Nature Communications, 4, 2013, 2487. https://doi.org/10.1038/ncomms3487

[441] Wu, Z.S., Parvez, K., Feng, X., Müllen, K., Journal of Materials Chemistry A, 2[22] 2014, 8288-8293. https://doi.org/10.1039/c4ta00958d

[442] Chang, J., Adhikari, S., Lee, T.H., Li, B., Yao, F., Pham, D.T., Le, V.T., Lee, Y.H., Advanced Energy Materials, 5[9] 2015, 1500003. https://doi.org/10.1002/aenm.201500003

[443] Song, B., Li, L., Lin, Z., Wu, Z.K., Moon, K.S., Wong, C.P., Nano Energy, 16, 2015, 470-478. https://doi.org/10.1016/j.nanoen.2015.06.020

[444] Xiao, H., Wu, Z.S., Chen, L., Zhou, F., Zheng, S., Ren, W., Cheng, H.M., Bao, X., ACS Nano, 11[7] 2017, 7284-7292. https://doi.org/10.1021/acsnano.7b03288

[445] Wen, F., Hao, C., Xiang, J., Wang, L., Hou, H., Su, Z., Hu, W., Liu, Z., Carbon, 75, 2014, 236-243. https://doi.org/10.1016/j.carbon.2014.03.058

[446] Amiri, M.H., Namdar, N., Mashayekhi, A., Ghasemi, F., Sanaee, Z., Mohajerzadeh, S., Journal of Nanoparticle Research, 18[8] 2016, 237. https://doi.org/10.1007/s11051-016-3552-5

[447] Zheng, B., Huang, T., Kou, L., Zhao, X., Gopalsamy, K., Gao, C., Journal of Materials Chemistry A, 2[25] 2014, 9736-9743. https://doi.org/10.1039/C4TA01868K

[448] Cai, W., Lai, T., Ye, J., Journal of Materials Chemistry A, 3[9] 2015, 5060-5066. https://doi.org/10.1039/C5TA00365B

[449] Aradilla, D., Delaunay, M., Sadki, S., Gérard, J.M., Bidan, G., Journal of Materials Chemistry A, 3[38] 2015, 19254-19262. https://doi.org/10.1039/C5TA04578A

[450] Li, J., Zhu, M., Wang, Z., Ono, T., Applied Physics Letters, 109[15] 2016, 153901. https://doi.org/10.1063/1.4964787

[451] Zhang, L., DeArmond, D., Alvarez, N.T., Malik, R., Oslin, N., McConnell, C., Adusei, P.K., Hsieh, Y.Y., Shanov, V., Small, 13[10] 2017, 1603114. https://doi.org/10.1002/smll.201603114

[452] Xie, B., Wang, Y., Lai, W., Lin, W., Lin, Z., Zhang, Z., Zou, P., Xu, Y., Zhou, S., Yang, C., Kang, F., Wong, C.P., Nano Energy, 26, 2016, 276-285. https://doi.org/10.1016/j.nanoen.2016.04.045

[453] Wang, S., Wu, Z.S., Zheng, S., Zhou, F., Sun, C., Cheng, H.M., Bao, X., ACS Nano, 11[4] 2017, 4283-4291. https://doi.org/10.1021/acsnano.7b01390

[454] Wu, G., Tan, P., Wu, X., Peng, L., Cheng, H., Wang, C.F., Chen, W., Yu, Z., Chen, S., Advanced Functional Materials, 27[36] 2017, 1702493. https://doi.org/10.1002/adfm.201702493

[455] Wu, X., Wu, G., Tan, P., Cheng, H., Hong, R., Wang, F., Chen, S., Journal of Materials Chemistry A, 6[19] 2018, 8940-8946. https://doi.org/10.1039/C7TA11135E

[456] Li, Q., Cheng, H., Wu, X., Wang, C.F., Wu, G., Chen, S., Journal of Materials Chemistry A, 6[29] 2018, 14112-14119. https://doi.org/10.1039/C8TA02124D

[457] Zhou, J., Chen, N., Ge, Y., Zhu, H., Feng, X., Liu, R., Ma, Y., Wang, L., Hou, W., Science China Materials, 61[2] 2018, 243-253. https://doi.org/10.1007/s40843-017-9168-9

[458] Zhai, S., Wang, C., Karahan, H.E., Wang, Y., Chen, X., Sui, X., Huang, Q., Liao, X., Wang, X., Chen, Y., Small, 14[29] 2018, 1800582. https://doi.org/10.1002/smll.201800582

[459] Naderi, L., Shahrokhian, S., Soavi, F., Journal of Materials Chemistry A, 8[37] 2020, 19588-19602. https://doi.org/10.1039/D0TA06561G

[460] Yuan, M., Luo, F., Rao, Y., Yu, J., Wang, Z., Li, H., Chen, X., Carbon, 183, 2021, 128-137. https://doi.org/10.1016/j.carbon.2021.07.014

[461] Shi, X., Wu, Z.S., Qin, J., Zheng, S., Wang, S., Zhou, F., Sun, C., Bao, X., Advanced Materials, 29[44] 2017, 1703034. https://doi.org/10.1002/adma.201703034

[462] Chi, K., Zhang, Z., Xi, J., Huang, Y., Xiao, F., Wang, S., Liu, Y., ACS Applied Materials and Interfaces, 6[18] 2014, 16312-16319. https://doi.org/10.1021/am504539k

[463] Couly, C., Alhabeb, M., Van Aken, K.L., Kurra, N., Gomes, L., Navarro-Suárez, A.M., Anasori, B., Alshareef, H.N., Gogotsi, Y., Advanced Electronic Materials, 4[1] 2018, 1700339. https://doi.org/10.1002/aelm.201700339

[464] Zhang, G., Liu, J., Ferroelectrics, 546[1] 2019, 67-73. https://doi.org/10.1080/00150193.2019.1592458

[465] Nagar, B., Dubal, D.P., Pires, L., Merkoçi, A., Gómez-Romero, P., ChemSusChem, 11[11] 2018, 1849-1856. https://doi.org/10.1002/cssc.201800426

[466] Ye, J., Tan, H., Wu, S., Ni, K., Pan, F., Liu, J., Tao, Z., Qu, Y., Ji, H., Simon, P., Zhu, Y., Advanced Materials, 30[27] 2018, 1801384. https://doi.org/10.1002/adma.201801384

[467] Li, Z., Cao, L., Qin, P., Liu, X., Chen, Z., Wang, L., Pan, D., Wu, M., Carbon, 139, 2018, 67-75. https://doi.org/10.1016/j.carbon.2018.06.042

[468] Meng, H., Zhang, L., Zhao, W., Zhou, Y., Zhang, Y., Zhang, G., Ferroelectrics, 581[1] 2021, 200-208. https://doi.org/10.1080/00150193.2021.1905743

[469] Zhang, L., Chen, Z., Zheng, S., Qin, S., Wang, J., Chen, C., Liu, D., Wang, L., Yang, G., Su, Y., Wu, Z.S., Bao, X., Razal, J., Lei, W., Journal of Materials Chemistry A, 7[23] 2019, 14328-14336. https://doi.org/10.1039/C9TA03620B

[470] Lee, H.U., Jin, J.H., Kim, S.W., Journal of Industrial and Engineering Chemistry, 71, 2019, 184-190. https://doi.org/10.1016/j.jiec.2018.11.021

[471] Bellani, S., Petroni, E., Del Rio Castillo, A.E., Curreli, N., Martín-García, B., Oropesa-Nuñez, R., Prato, M., Bonaccorso, F., Advanced Functional Materials, 29[14] 2019, 1807659. https://doi.org/10.1002/adfm.201807659

[472] Li, B., Hu, N., Su, Y., Yang, Z., Shao, F., Li, G., Zhang, C., Zhang, Y., ACS Applied Materials and Interfaces, 11[49] 2019, 6044-46053. https://doi.org/10.1021/acsami.9b12225

[473] Delekta, S.S., Laurila, M.M., Mäntysalo, M., Li, J., Nano-Micro Letters, 12[1] 2020, 40.

[474] Bhattacharya, G., Fishlock, S.J., Pritam, A., Sinha Roy, S., McLaughlin, J.A., Advanced Sustainable Systems, 4[4] 2020, 1900133. https://doi.org/10.1002/adsu.201900133

[475] Li, F., Qu, J., Li, Y., Wang, J., Zhu, M., Liu, L., Ge, J., Duan, S., Li, T., Bandari, V.K., Huang, M., Zhu, F., Schmidt, O.G., Advanced Science, 7[19] 2020, 2001561. https://doi.org/10.1002/advs.202001561

[476] Chen, H., Chen, S., Zhang, Y., Ren, H., Hu, X., Bai, Y., ACS Applied Materials and Interfaces, 12[50] 2020, 56319-56329. https://doi.org/10.1021/acsami.0c16976

[477] Chen, Y., Guo, M., Xu, L., Cai, Y., Tian, X., Liao, X., Wang, Z., Meng, J., Hong, X., Mai, L., Nano Research. 15, 2022, 1492-1499. https://doi.org/10.1007/s12274-021-3693-4

[478] Feng, X., Ning, J., Wang, D., Zhang, J., Dong, J., Zhang, C., Shen, X., Hao, Y.,

Journal of Power Sources, 418, 2019, 130-137.
https://doi.org/10.1016/j.jpowsour.2019.01.093

[479] Zhang, G., Liu, C., Liu, L., Li, X., Liu, F., Ferroelectrics, 547[1] 2019, 129-136.
https://doi.org/10.1080/00150193.2019.1592492

[480] He, Y., Zhang, P., Wang, M., Wang, F., Tan, D., Li, Y., Zhuang, X., Zhang, F.,
Feng, X., Materials Horizons, 6[5] 2019, 1041-1049.
https://doi.org/10.1039/C9MH00063A

[481] Du, J., Mu, X., Zhao, Y., Zhang, Y., Zhang, S., Huang, B., Sheng, Y., Xie, Y.,
Zhang, Z., Xie, E., Nanoscale, 11[30] 2019, 14392-14399.
https://doi.org/10.1039/C9NR03917A

[482] Shi, X., Tian, L., Wang, S., Wen, P., Su, M., Xiao, H., Das, P., Zhou, F., Liu, Z.,
Sun, C., Wu, Z.S., Bao, X., Journal of Energy Chemistry, 52, 2020, 284-290.
https://doi.org/10.1016/j.jechem.2020.04.064

[483] Esfahani, M.Z., Khosravi, M., Journal of Power Sources, 462, 2020, 228166.
https://doi.org/10.1016/j.jpowsour.2020.228166

[484] Zhang, L.D., Liu, C.F., Li, X., Liu, F.S., Zhao, W.Q., Zhang, G.Y., Ferroelectrics,
564[1] 2020, 146-152. https://doi.org/10.1080/00150193.2020.1761709

[485] Wu, Y., Zhang, Y., Liu, Y., Cui, P., Chen, S., Zhang, Z., Fu, J., Xie, E., ACS
Applied Materials and Interfaces, 12[38] 2020, 42933-42941.
https://doi.org/10.1021/acsami.0c11085

[486] Zhao, Y., Du, J., Li, Y., Li, X., Zhang, C., Zhang, X., Zhang, Z., Zhou, J., Pan, X.,
Xie, E., ACS Applied Energy Materials, 3[9] 2020, 8415-8422.
https://doi.org/10.1021/acsaem.0c01036

[487] Liu, H., Moon, K.S., Li, J., Xie, Y., Liu, J., Sun, Z., Lu, L., Tang, Y., Wong, C.P.,
Nano Energy, 77, 2020, 105058. https://doi.org/10.1016/j.nanoen.2020.105058

[488] Yu, X., Li, N., Zhang, S., Liu, C., Chen, L., Han, S., Song, Y., Han, M., Wang, Z.,
Journal of Power Sources, 478, 2020, 229075.
https://doi.org/10.1016/j.jpowsour.2020.229075

[489] Yuan, Y., Jiang, L., Li, X., Zuo, P., Xu, C., Tian, M., Zhang, X., Wang, S., Lu, B.,
Shao, C., Zhao, B., Zhang, J., Qu, L., Cui, T., Nature Communications, 11[1]
2020, 6185. https://doi.org/10.1038/s41467-020-19985-2

[490] Zhang, C., Peng, Z., Huang, C., Zhang, B., Xing, C., Chen, H., Cheng, H., Wang,
J., Tang, S., Nano Energy, 81, 2021, 105609.

https://doi.org/10.1016/j.nanoen.2020.105609

[491] Peng, Z., Jia, J., Ding, H., Yu, H., Shen, Y., Zhang, J., Tao, W., Zhang, C., Wang, J., Cheng, H., Journal of Materials Research, 36[9] 2021, 1927-1936. https://doi.org/10.1557/s43578-021-00228-z

[492] Peng, C., Li, Q., Niu, L., Yuan, H., Xu, J., Yang, Q., Yang, Y., Li, G., Zhu, Y., Carbon, 175, 2021, 27-35. https://doi.org/10.1016/j.carbon.2020.12.089

[493] Liu, L., Lu, J.Y., Long, X.L., Zhou, R., Liu, Y.Q., Wu, Y.T., Yan, K.W., Science China Technological Sciences, 64[5] 2021, 1065-1073. https://doi.org/10.1007/s11431-020-1763-5

[494] Bouzina, A., Perrot, H., Sel, O., Debiemme-Chouvy, C., ACS Applied Nano Materials, 4[5] 2021, 4964-4973. https://doi.org/10.1021/acsanm.1c00489

www.ingramcontent.com/pod-product-compliance
Lightning Source LLC
Chambersburg PA
CBHW071650210326
41597CB00017B/2173